T0295176

Basic Electronics: Devices and Systems

Basic Electronics: Devices and Systems

Charlotte Green

www.statesacademicpress.com

States Academic Press,
109 South 5th Street,
Brooklyn, NY 11249, USA

Visit us on the World Wide Web at:
www.statesacademicpress.com

ISBN: 978-1-63989-717-9

Cataloging-in-publication Data

Basic electronics : devices and systems / Charlotte Green.
p. cm.
Includes bibliographical references and index.
ISBN 978-1-63989-717-9
1. Electronics. 2. Electrical engineering. I. Green, Charlotte.
TK7803 .B37 2023
621.381--dc23

Table of Contents

Preface

Electronic devices are the devices which work on electronic variables such as power, voltage or current. These systems are used for controlling the flow of electrical currents for information processing and system control. Electronic devices contain two types of components, namely, passive components and active components. Passive components are without gain or directionality such as resistors, capacitors, diodes and inductors, whereas active components are those having gain or directionality. Active components include transistors, integrated circuits (ICs), and logic gates; whereas circuit is a passive component. Electronic systems are created to process electrical signals. The electronic systems can have a number of inputs and outputs. Some examples of an electronic system are an audio system, MP3 player and television. Electronic devices and systems are utilized for the acquisition or acceptance, processing, storage, display, analysis, protection, disposition, and transfer of information. This book outlines the applications of basic electronics devices and systems in detail. Coherent flow of topics, student-friendly language, and extensive use of examples make this book an invaluable source of knowledge.

This book has been the outcome of endless efforts put in by authors and researchers on various issues and topics within the field. The book is a comprehensive collection of significant researches that are addressed in a variety of chapters. It will surely enhance the knowledge of the field among readers across the globe.

It gives us an immense pleasure to thank our researchers and authors for their efforts to submit their piece of writing before the deadlines. Finally in the end, I would like to thank my family and colleagues who have been a great source of inspiration and support.

Charlotte Green

1

Review of Semiconductor Physics

1.1 Classification of Metals, Mobility and Conductivity

By considering their electrical properties, we can divide the materials into three groups namely insulator, semiconductor and conductor.

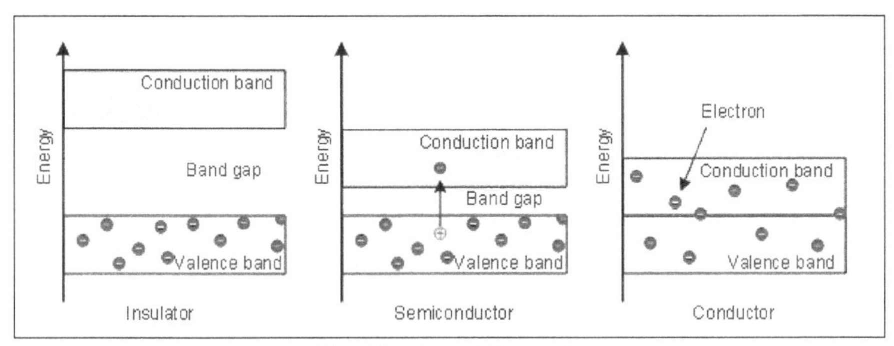

Classification of Materials.

Conductors	They have a lot of free electrons which flow through the material. Example: All metals and semi-metals like carbon-graphite, antimony and arsenic.
Insulators	They have a very few free electrons. Examples: Plastic, glass and wood.
Semiconductors	They lie between the extremes of good conductors and good insulators. They are crystalline materials which are insulators when pure but they will conduct when the impurity is added in response to light, heat, voltage, etc., Example: Elements like silicon (Si), germanium (Ge), selenium (Se) and compounds like gallium arsenide (GaAs) and indium antimonide (InSb).

Concepts of Effective Mass of Electron: Concept of Hole

Effective mass is given as,

$$\left(\frac{1}{m^*}\right)_{\mu\nu} = \frac{1}{\hbar^2}\frac{d^2E\left(\vec{k}\right)}{dk_\mu dk_\nu}$$

In a solid, the electron (hole) effective mass represents the movement of electrons in an applied field. The effective mass reflects the inverse of the band curvature. If m* is lower, then the curvature is larger.

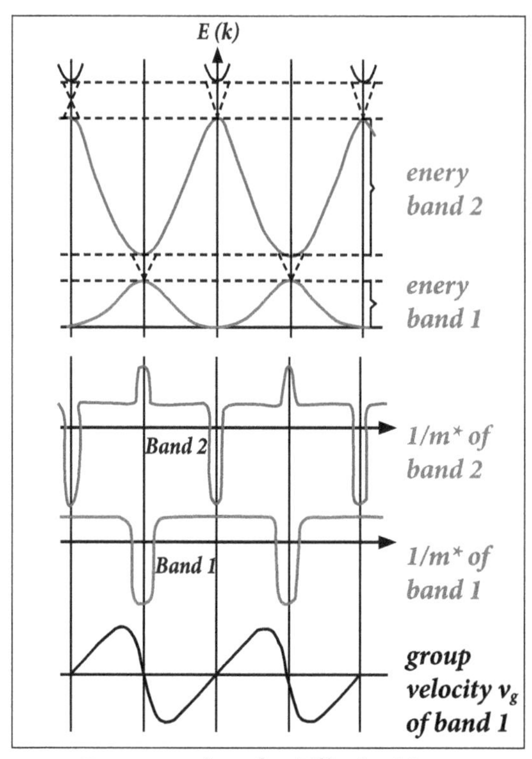

Representation of an Effective Mass.

Flatter bands have larger effective mass where electrons are more accelerated. In a three dimensional band structure, the effective mass is a tensor.

1.1.1 Mobility and Conductivity

Mobility in solid-state physics is a measurement of a particular type of charged particle which moves through a solid material under the influence of an electric field. Such particles are pulled by the electric field and they periodically collide with atoms of the solid.

This combination of the electric field and collisions causes the particles to move with an average velocity known as the drift velocity where the charge carrier is a negatively charged electron. Mobility is defined as the value of the drift velocity per unit of electric field strength. Mobility will be larger when the particle moves faster at a given electric field strength. The mobility of a particular type of particle in a given solid varies with temperature.

It depends on the type of solid. For example, in semiconductors, electric current is carried by the motion of positively charged particles called holes each of which corresponds to the absence of an electron where the determination of their separate mobilities is difficult. Many electronic devices require high mobilities for efficient operation.

The electrical conductivity depends on the negative exponential of band gap (E_g) between the valance band and conduction band and also the mobilities of both holes and electrons. The mobilities in a pure semiconductor is determined by the interaction of electron with lattice waves or phonons.

1.2 Electrons and Holes in an Intrinsic Semiconductor

Intrinsic semiconductors: A pure semiconductor without any impurities is known as an intrinsic semiconductor.

Example: Ge, Si (Pure form)

Carrier Concentration in an Intrinsic Semiconductor

In a semiconductor, both electrons and holes are charge carriers (known as carrier concentration). A semiconductor in which holes and electrons are created by thermal excitation across the energy gap is called as an intrinsic semiconductor.

In an intrinsic semiconductor, the number of holes is equal to the number of free electrons.

At T = 0 K, valence band is completely filled and conduction band is completely empty. Thus, the intrinsic semiconductor behaves as a perfect insulator.

At T > 0 K, the electron from the valence band is shifted to the conduction band across the band gap.

Thus, there are number of free electrons and holes in an intrinsic semiconductor. Fermi level lies in the midway between the conduction band and valance band in an intrinsic semiconductors.

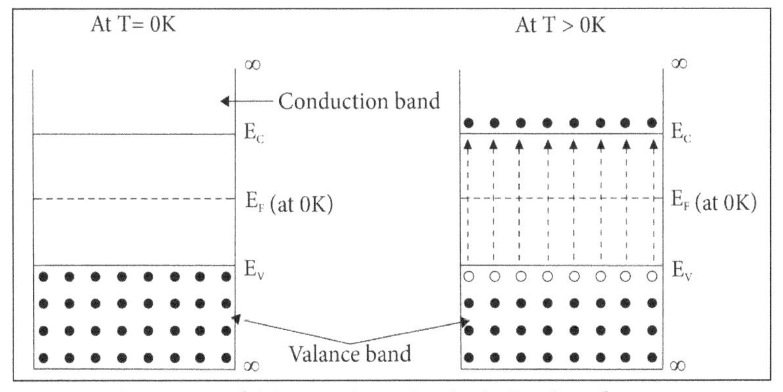

Energy Band Diagram in an Intrinsic Semiconductor.

Density of electrons in conduction band is given as,

$$N_e = N_e = 2\left[\frac{\left(2\pi m_e^* K_B T\right)}{h^2}\right]^{3/2} \exp\left[\frac{E_F - E_C}{K_B T}\right]$$

Density of holes in valence band is given as,

$$N_h = N_h = 2\left[\frac{\left(2\pi m_e^* K_B T\right)}{h^2}\right]^{3/2} \exp\left[\frac{E_v - E_F}{K_B T}\right]$$

Intrinsic Carrier Concentration

In an intrinsic semiconductor,

$$N_e = N_h = n_i$$

Where,

n_i – Intrinsic carrier concentration

$$n_i^2 = N_e N_h$$

$$n_i^2 = 2\left[\frac{2\pi m_e^* K_B T}{h^2}\right]^{3/2} \exp\left[\frac{E_F - E_C}{K_B T}\right] 2\left[\frac{2\pi m_h^* K_B T}{h^2}\right]^{3/2} \exp\left[\frac{E_V - E_F}{K_B T}\right]$$

$$n_i^2 = 4\left[\left[\frac{2\pi K_B T}{h^2}\right]^{3/2}\right]^2 \left[m_e^* m_h^*\right]^{3/2} \exp\left[\frac{-[E_C - E_V]}{K_B T}\right]$$

$$n_i^2 = 4\left[\left[\frac{2\pi K_B T}{h^2}\right]^{3/2}\right]^2 \left[m_e^* m_h^*\right]^{3/2} \exp\left[\frac{-E_g}{K_B T}\right]$$

$$n_i = 4\left[4\left[\left[\frac{2\pi K_B T}{h^2}\right]^{3/2}\right]^2 \left[m_e^* m_h^*\right]^{3/2} \exp\left[\frac{-E_g}{K_B T}\right]\right]^{1/2}$$

$$n_i = 2\left(\frac{2\pi K_B T}{h^2}\right)^{3/2} \left[m_e^* m_h^*\right]^{3/4} \exp\left(\frac{-E_g}{K_B T}\right)^{1/2}$$

$$n_i = 1\left[\frac{2\pi K_B T}{h^2}\right]^{3/2}\left[m_e^* m_h^*\right]^{3/4}\exp\left[\frac{-E_g}{2K_B T}\right]$$

Where,

$E_g = E_c - E_v$ is the energy gap between conduction band and the valence band.

The above equation is called as an intrinsic carrier concentration.

For an intrinsic semiconductor, even if impurity is added to increase N_e, there will be decrease in N_h and hence the product $N_e N_h$ remains constant. This is called as the law of mass action.

Expression for Electrical Conductivity in an Intrinsic Semiconductor

Expression for the electrical conductivity is given as,

$$\sigma = n\,e\,\mu$$

The intrinsic electrical conductivity is given as,

$$\sigma_i = \left[n\,e\,\mu_e + p\,e\,\mu_h\right]$$

$$n = p = n_i$$

$$\sigma_i = \left[n_i\,e\,\mu_e + n_i\,e\,\mu_h\right]$$

$$\sigma_i = n_i\,e\left[\mu_e + \mu_h\right]$$

Where,

μ_e – Electron mobility

μ_h – Hole mobility.

$$\sigma_i = \left[\mu_e + \mu_h\right]e\,2\left[\frac{2\pi K_B T}{h^2}\right]^{3/2}\left[m_e^* m_h^*\right]^{3/4}\exp\left[-\frac{E_g}{2K_B T}\right]$$

The electrical conductivity depends on the negative exponential of band gap (E_g) between the valance band and conduction band and also for the mobilities of both holes and electrons. The mobilities in a pure semiconductor is due to the interaction of electron with lattice waves or phonons.

If we can neglect $[\mu_e + \mu_h]$ in the above equation, we have,

The electrical conductivity is given as,

$$\sigma_i = C \exp\left[-\frac{E_g}{2K_BT}\right] \quad ...(1)$$

Where,

C is a constant

Taking log on both sides of equation (1), we have,

$$\log \sigma_i = \text{Log}\left[C \exp\left(-\frac{E_g}{2K_BT}\right)\right]$$

$$\log \sigma_i = \text{Log } C + \left(\frac{-E_g}{2K_BT}\right)$$

$$\log \sigma_i = \text{Log } C - \left(\frac{E_g}{2K_BT}\right)$$

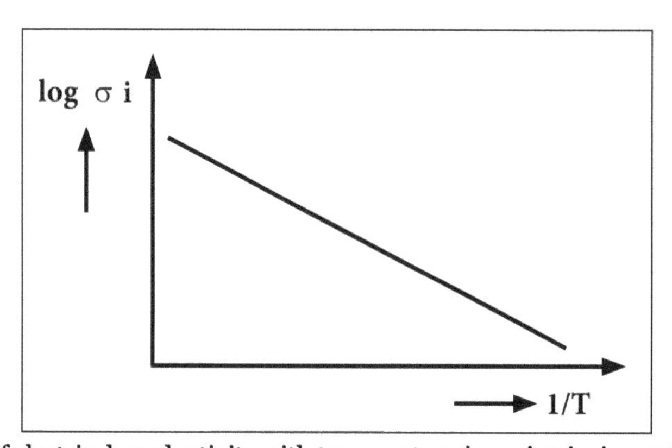

Variation of electrical conductivity with temperature in an intrinsic semiconductor.

A graph is drawn between 1/T and log σ_i. From the above graph, it is noted that this electrical conductivity increases with temperature.

Determination of Band Gap

Electrical conductivity is given as,

$$\sigma_i = C \exp\left[\frac{-E_g}{2K_BT}\right]$$

Resistivity is given as,

$$\rho_i = \frac{1}{\sigma_i}$$

$$\rho_i = \frac{1}{C}\exp\left(\frac{E_g}{2K_BT}\right) \qquad ...(1)$$

Resistivity is defined as the resistance per unit area per unit length,

$$\rho_i = \frac{R_iA}{L} \qquad ...(2)$$

ρ_i – Resistivity

A - Cross sectional area

L – Length

Equating equation (1) and (2), we have,

$$\frac{R_iA}{L} = \frac{1}{C}\exp\left[\frac{E_g}{2K_BT}\right]$$

$$R_i = \frac{L}{AC}\exp\left[\frac{E_g}{2K_BT}\right]$$

Taking log on both sides, we have,

$$\log R_i = \log\frac{L}{AC} + \left[\frac{E_g}{2K_BT}\right]$$

The above equation gives us a method of determining the energy gap of an intrinsic material. If we find the resistance of an intrinsic semiconductor using post office box or carry Foster's bridge at various temperatures, we can plot a graph between 1/T and log R_i.

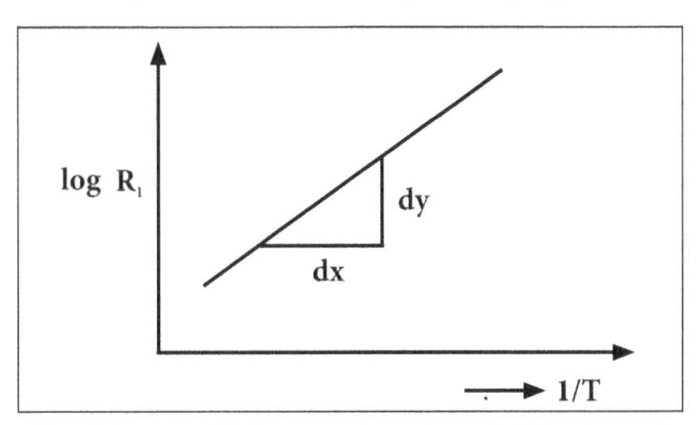

Variation of resistance with temperature in an intrinsic semiconductor.

From the curve, we have,

$$\frac{dy}{dx} = \frac{E_g}{2K_B T}$$

Therefore by finding the slope of line, the energy band gap is given as,

$$E_g = 2K_B \left(\frac{dy}{dx}\right) \text{Joules}$$

1.3 Extrinsic Semiconductor, Hall Effect, Generation and Recombination of Charges, Diffusion, Continuity Equation, Injected Minority Carriers and Law of Junction

Extrinsic Semiconductor

Compound semiconductor is a semiconductor that consists of elements from two or more different groups of the periodic table. They also have known as direct band gap semiconductors.

i.e., III - V group, II - VI group and IV - VI group.

Here the recombination of electron and hole takes place directly during recombination where photons are emitted.

Example: GaAs, GaP

Based on the type of impurity, they are classified into two types as follows,

- N-type semiconductor.
- P-type semiconductor.

The band gap represents the minimum energy difference between the top of the valence band and the bottom of the conduction band. However, the top of the valence band and the bottom of the conduction band are not at the same value of the electron momentum.

- In a direct band gap semiconductor, the top of the valence band and the bottom of the conduction band occur at the same value of momentum.
- In an indirect band gap semiconductor, the value of momentum is maximum at valence band and minimum at the conduction band.

Carrier Concentration in p-type Semiconductor

p-type Semiconductor

If trivalent (Aluminum, Gallium, Indium) impurities are doped with pure semiconducting material, then holes are produced which is called as p - type semiconductor.

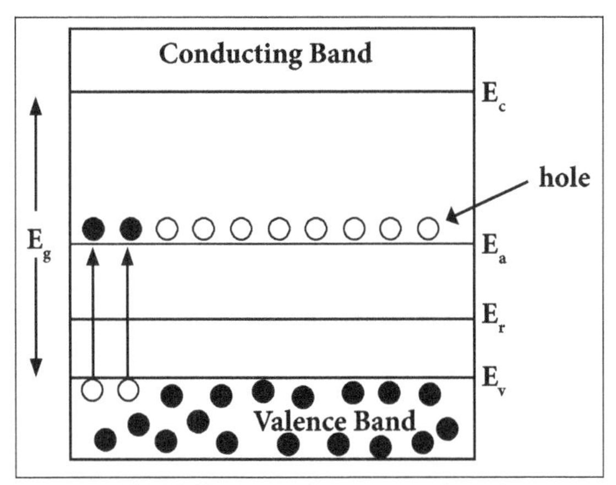

Energy level diagram of p-type semiconductor.

Density of holes in the valence band in an intrinsic semiconductors is given as,

$$N_h = 2\left[\frac{2\pi m_e^* K_B T}{h^2}\right]^{3/2} \exp\left[\frac{E_V - E_F}{K_B T}\right] \quad ...(1)$$

$$N_V = 2\left[\frac{2\pi m_h^* K_B T}{h^2}\right]^{3/2} \quad ...(2)$$

Density of holes is given as,

$$N_h = N_V \exp\left[\frac{E_V - E_F}{K_B T}\right] \quad ...(3)$$

Density of ionized acceptor atoms is given as,

$$N_A F[E_A] = N_A \exp\left[\frac{E_F - E_A}{K_B T}\right] \quad ...(4)$$

At equilibrium condition,

Number of holes per unit volume in valence band equation = Number of electron per unit volume in acceptor energy level equation,

$$N_V \exp\left[\frac{E_V - E_F}{K_B T}\right] = N_A \exp\left[\frac{E_F - E_A}{K_B T}\right]$$

$$\exp\left[\frac{E_V - E_F}{K_B T}\right]\exp\left[\frac{-[E_F - E_A]}{K_B T}\right] = \frac{N_A}{N_V}$$

$$\exp\left[\frac{E_V - E_F - E_F + E_A}{K_B T}\right] = \frac{N_A}{N_V}$$

Taking log on both sides, we have,

Substituting the value N_v, we have,

$$\log\left[\exp\left[\frac{E_V - E_F - E_F + E_A}{K_B T}\right]\right] = \log\left[\frac{N_A}{N_v}\right]$$

$$\left[\frac{E_v - E_F - E_F + E_A}{K_B T}\right] = \log\left[\frac{N_A}{N_v}\right]$$

$$E_v - E_F - E_F + E_A = K_B T \log\left[\frac{N_A}{N_v}\right]$$

$$-2E_F = -[E_v + E_A] + K_B T \log\left[\frac{N_A}{N_v}\right]$$

$$E_F = \frac{(E_v + E_A)}{2} - \frac{K_B T}{2}\log\left[\frac{N_A}{N_v}\right]$$

At T = 0K,

$$E_F = \frac{(E_v + E_A)}{2} - \frac{K_B T}{2}\log\left[\frac{N_A}{2\left(\frac{2\pi m_h^* K_B T}{h^2}\right)^{3/2}}\right] \quad \ldots(5)$$

At 0K, fermi level in p-type semiconductor lies exactly at the middle of the acceptor level and at the top of the valance band.

$$E_F = \left[\frac{(E_v + E_A)}{2}\right]$$

Expression for the Density of Holes in Valence Band in Terms of N_A

When the temperature increases, the acceptor atoms are ionized. Further increase in temperature results in the generation of electron hole pairs due to the breaking of covalent bonds where materials tends to exhibit the properties of an intrinsic semiconductor and the fermi level moves towards the intrinsic fermi level.

Density of holes in valence band is given as,

$$N_h = 2\left[\frac{\left(2\pi m_h^* K_B T\right)}{h^2}\right]^{3/2} \exp\left[\left(\frac{E_v - E_F}{K_B T}\right)\right]$$

Substituting equation (5) in the above equation, we have,

$$N_h = 2\left(\frac{2\pi m_h^* K_B T}{h^2}\right)^{\frac{3}{2}} \exp - \frac{K_B T}{2}\log\left[\frac{E_v - \left[\left(\frac{(E_v - E_A)}{2}\right) - \frac{K_B T}{2}\log\frac{N_A}{2\left(\frac{2\pi m_h^* K_B T}{h^2}\right)^{1/2}}\right]}{K_B T}\right]$$

$$= 2\left[\frac{\left(2\pi m_h^* K_B T\right)}{h^2}\right]^{3/2} \exp\left[\left(\frac{E_v - E_A}{2K_B T}\right) + \frac{1}{2}\log\left(\frac{N_A}{2\left[\frac{\left[2\pi m_h^* K_B T\right]}{h^2}\right]^{3/2}}\right)\right]$$

$$= 2\left[\frac{\left(2\pi m_h^* K_B T\right)}{h^2}\right]^{3/2} \exp\left[\left(\frac{E_v - E_A}{2K_B T}\right) + \log\left(\frac{N_A}{2\left[\frac{\left[2\pi m_h^* K_B T\right]}{h^2}\right]^{3/2}}\right)^{1/2}\right]$$

$$= 2\left[\frac{2\pi m_h^* K_B T}{h^2}\right]^{3/2} \exp\left[\left(\frac{E_v - E_A}{2K_B T}\right) + \log\left(\frac{N_A^{1/2}}{2^{1/2}\left[\frac{\left[2\pi m_h^* K_B T\right]}{h^2}\right]^{3/4}}\right)\right]$$

$$= 2\left[\frac{\left(2\pi m_h^* K_B T\right)}{h^2}\right]^{3/2} \exp\left(\frac{E_v - E_A}{2K_B T}\right)\left(\frac{N_A^{1/2}}{2^{1/2}\left[\frac{2\pi m_h^* K_B T}{h^2}\right]^{3/4}}\right)$$

$$(2N_A)^{1/2} \exp\left[\frac{E_v - E_A}{2K_BT}\right]\left[\frac{(2\pi m_h^* K_BT)}{h^2}\right]^{3/2}\left[\frac{(2\pi m_h^* K_BT)}{h^2}\right]^{-3/4}$$

$$= (2N_A)^{1/2} \exp\left[\frac{E_v - E_A}{2K_BT}\right]\left[\frac{(2\pi m_h^* K_BT)}{h^2}\right]^{3/4}$$

$$E_A - E_V = \Delta E$$

Where,

ΔE is known as ionization energy of acceptors.

i.e. ΔE represents the energy required for an electron to move from valance band (E_V) to an acceptor energy level (E_A).

$$N_h = (2N_A)^{1/2} \exp\left[\frac{-\Delta E}{2K_BT}\right]\left[\frac{(2\pi m_h^* K_BT)}{h^2}\right]^{3/4}$$

Carrier Concentration in n-type Semiconductor

If pentavalent (Phosphorous, Arsenic, Antimony) impurities are doped with pure semi-conducting material, then the free electrons are produced which is called as n-type semiconductor.

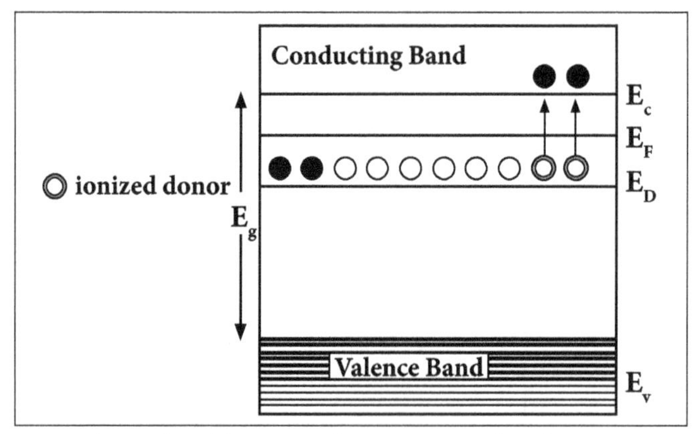

Energy level diagram of an n-type semiconductor.

Density of electrons in conduction band in an intrinsic semiconductor is given as,

$$N_e = \left[2\left[\frac{2\pi m_e^* K_BT}{h^2}\right]^{3/2} \exp\left[\frac{E_F - E_C}{K_BT}\right]\right] \quad ...(1)$$

$$N_c = 2\left[\frac{2\pi m_e^* K_B T}{h^2}\right]^{3/2}$$

Density of electrons is given as,

$$N_e = N_C \exp\left[\frac{E_F - E_C}{K_B T}\right] \qquad ...(2)$$

Density of ionized donor atoms is given as,

$$N_D\left[1 - F(E_D)\right] = N_D \exp\left[\frac{E_D - E_F}{K_B T}\right] \qquad ...(3)$$

At equilibrium condition,

Number of electrons per unit volume in conduction band equation = Number of holes per unit volume in donor energy level,

$$N_C \exp\left[\frac{E_F - E_C}{K_B T}\right] = N_D \exp\left[\frac{E_D - E_F}{K_B T}\right]$$

$$\frac{\exp\left[\frac{E_F - E_C}{K_B T}\right]}{\exp\left[\frac{E_D - E_F}{K_B T}\right]} = \frac{N_D}{N_C}$$

$$\exp\left[\frac{E_F - E_C - E_D + E_F}{K_B T}\right] = \frac{N_D}{N_C}$$

Taking log on both sides, we have,

$$\log\left[\exp\left[\frac{E_F - E_C - E_D + E_F}{K_B T}\right]\right] = \log\left[\frac{N_D}{N_C}\right]$$

$$\left[\frac{E_F - E_C - E_D + E_F}{K_B T}\right] = \log\left[\frac{N_D}{N_C}\right]$$

$$2E_F = (E_C + E_D) + K_B T \log\left[\frac{N_D}{N_C}\right]$$

$$E_F = \frac{(E_C + E_D)}{2} + \frac{K_B T}{2}\log\left[\frac{N_D}{N_C}\right] \qquad ...(4)$$

At T = oK,

$$E_F = \left[\frac{(E_C + E_D)}{2}\right] \quad ...(5)$$

At T = oK, the fermi level in n-type semiconductor lies exactly in middle of the conduction level (E_C) and donor level (E_D).

The above equation shows that the electron concentration in the conduction band is proportional to the square root of the donor concentration.

Expression for the Density of Electrons in Conduction Band in Terms of N_D

When the temperature increases, the donor atoms are ionized and the fermi level drops. For a particular temperature, all donor atoms are ionized and further increase in temperature results in the generation of electron hole pairs due to the breaking of covalent bonds and materials tends to behave in an intrinsic manner.

Density of electrons in conduction band is given as,

$$N_e = 2\left[\frac{(2\pi m_e^* K_B T)}{h^2}\right]^{3/2} \exp\left[\frac{E_F - E_C}{K_B T}\right]$$

Substituting equation (4) in the above equation, we have,

$$N_e = 2\left[\frac{(2\pi m_e^* K_B T)}{h^2}\right]^{3/2} \exp\left[\frac{\left(\frac{E_C + E_D}{2}\right) + \frac{K_B T}{2}\log\left(\frac{N_D}{2\left[\left[\frac{2\pi m_e^* K_B T}{h^2}\right]\right]}\right) - E_C}{K_B T}\right]$$

$$= 2\left[\frac{(2\pi m_e^* K_B T)}{h^2}\right]^{3/2} \exp\left[\left(\frac{E_C + E_D}{2}\right) + \frac{1}{2}\log\left(\frac{N_D}{2\left[\left[\frac{2\pi m_e^* K_B T}{h^2}\right]\right]^{3/2}}\right) - \frac{E}{K_B T}\right]$$

$$= 2\left[\frac{2\left(2\pi m_e^* K_B T\right)}{h^2}\right]^{3/2} \exp\left[\left(\frac{E_D - E_C}{2K_B T}\right) + \log\left(\frac{N_D}{2\left[\frac{\left[2\pi m_e^* K_B T\right]}{h^2}\right]^{3/2}}\right)^{1/2}\right]$$

$$= 2\left[\frac{2\left(2\pi m_e^* K_B T\right)}{h^2}\right]^{3/2} \exp\left[\left(\frac{E_D - E_C}{2K_B T}\right) + \log\left(\frac{N_D^{1/2}}{2^{1/2}\left[\frac{\left[2\pi m_e^* K_B T\right]}{h^2}\right]^{3/2}}\right)\right]$$

$$= 2\left[\frac{2\left(2\pi m_e^* K_B T\right)}{h^2}\right]^{3/2} \exp\left[\left(\frac{E_D - E_C}{2K_B T}\right)\left(\frac{N_D^{1/2}}{2^{1/2}\left[\frac{\left[2\pi m_e^* K_B T\right]}{h^2}\right]^{3/2}}\right)\right]$$

$$= \left(2N_D\right)^{1/2} \exp\left[\frac{E_D - E_C}{2K_B T}\right]\left[\frac{\left(2\pi m_e^* K_B T\right)}{h^2}\right]^{3/2}\left[\frac{\left(2\pi m_e^* K_B T\right)}{h^2}\right]^{-3/4}$$

$$= \left(2N_D\right)^{1/2} \exp\left[\frac{E_D - E_C}{2K_B T}\right]\left[\frac{\left(2\pi m_e^* K_B T\right)}{h^2}\right]^{3/4}$$

Here, $E_C - E_D = \Delta E$ is known as ionization energy of donors i.e. ΔE represents the amount of energy required to transfer an electron from donor energy level $\left(E_D\right)$ to conduction band $\left(E_C\right)$,

$$\boxed{N_e = \left(2N_D\right)^{1/2} \exp\left[\frac{-\Delta E}{2K_B T}\right]\left[\frac{\left(2\pi m_e^* K_B T\right)}{h^2}\right]^{3/4}}$$

1.3.1 Hall Effect in p-type and n-type Semiconductor

When conductor carrying a current is placed in a transverse magnetic field, an electric field is produced inside the conductor in a direction normal to both the current and the magnetic field. This phenomenon is known as Hall Effect and the generated voltage is called as the hall voltage.

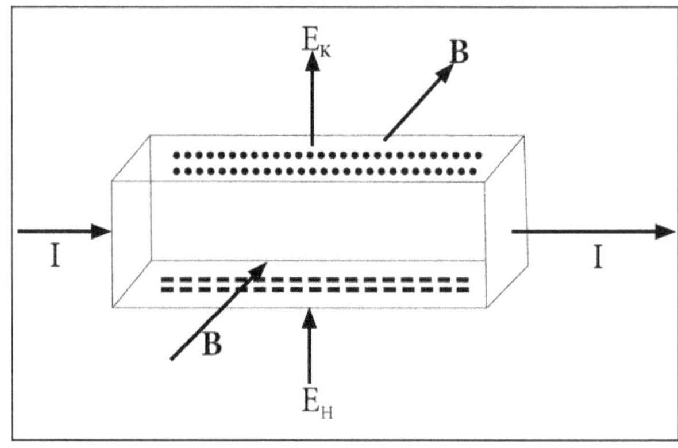

Hall Effect.

Hall Effect in n-type Semiconductor

Let us consider an n-type material in which the current is allowed to pass along x-direction from left to right (electrons move from right to left) and the magnetic field is applied in z-directions which produces hall voltage in y-direction.

Since the direction of current is from left to right, the electrons moves from right to left in x-direction as shown in the below figure.

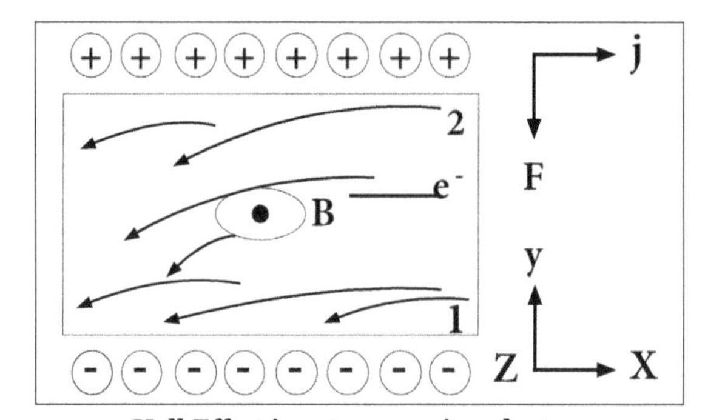

Hall Effect in n-type semiconductor.

Due to the applied magnetic field, the electrons move towards downward direction with the velocity 'v' and the negative charge accumulates at face (1) of the material as shown in the above figure.

Therefore, a potential difference is established between face (2) and face (1) which gives rise to field E_H in the negative y-direction.

Force due to potential difference $= -\ eE_H$ ...(1)

Force due to magnetic field $= -Bev$ ...(2)

At equilibrium,

Equation (1) = Equation (2)

$$-eE_H = -Bev$$

$$E_H = Bv \qquad ...(3)$$

Current density J_x in the x direction is given as,

$$J_x = -n_e ev$$

$$V = \frac{J_X}{N_e e} \qquad ...(4)$$

Substituting equation (4) in equation (3), we have,

$$E_H = \frac{BJ_X}{n_e e} \qquad ...(5)$$

$$E_H = R_H J_x B \quad ...(6)$$

Where,

$$R_H = \text{Hall coefficient}$$

$$R_H = -(1/n_e e) \qquad ...(7)$$

The negative sign indicates that the field is developed in the negative y-direction.

Hall Effect in p-type Semiconductor

Let us consider a p-type material for which the current is passed along x-direction from left to right and magnetic field is applied along z-direction as shown in the below figure. Since the direction of current is from left to right, the holes will move in the same direction.

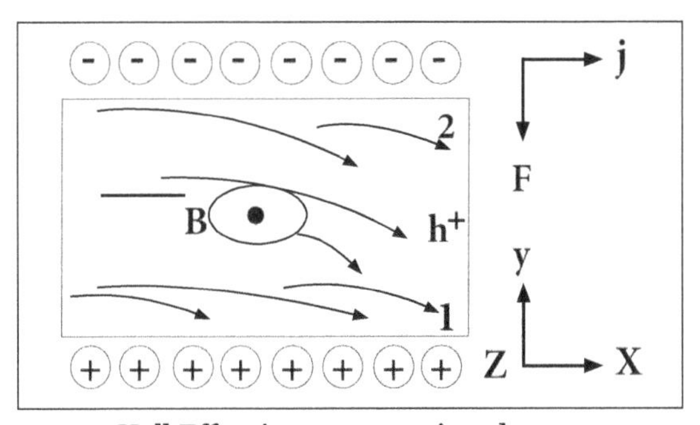

Hall Effect in p-type semiconductor.

Due to the applied magnetic field, the holes move towards the downward direction with velocity 'v' and accumulate at the face (1) as shown in the above figure.

A potential difference is established between face (1) and (2) in the positive y-direction.

Force due to the potential difference $= eE_H$ ...(8)

At equilibrium,

Equation (7) = Equation (8)

$$eE_H = Bev$$

$$E_H = Bv$$

Current density is given as,

$$J_x = n_h ev \quad ...(9)$$

$$V = \frac{J_X}{n_h}e \quad ...(10)$$

Where,

n_h = Hole density

Substituting equation (10) in (9), we have,

$$E_H = \frac{BJ_X}{n_h e}$$

$$E_H = R_H J_x B \quad ...(11)$$

Where,

$$R_H = \frac{1}{n_h e}$$

The above equation (11) represents the hall coefficient and the positive sign indicates that the hall field is developed in the positive y-direction.

Experimental Determination of Hall Effect

A semiconductor slab of thickness 't' and breadth 'b' is taken and current is passed using the battery as

Shown in the figure.

The slab is placed between the pole of an electromagnet so that current direction coincides with x-axis and magnetic field coincides with z-axis. The hall voltage (V_H) is measured by placing two probes at the center of the top and bottom faces of the slab (y-axis).

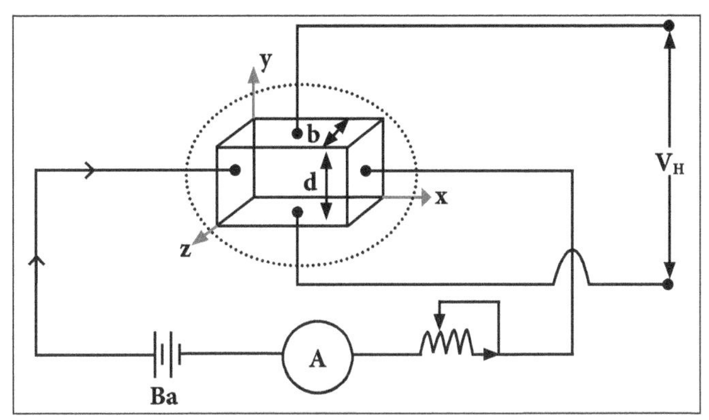

Experimental Setup For Hall Effect.

If B is the applied magnetic field and V_H is the hall voltage, then the hall coefficient is given as,

$$R_H = \frac{V_H b}{I_X B} \quad ...(1)$$

Mobility of Charge Carriers

Hall co-efficient is given as,

$$R_H = \frac{-1}{ne} \quad ...(2)$$

The above expression is valid only for conductors where the velocity is taken as the drift velocity. But for semiconductors, velocity is taken as average velocity. Hence, RH for an 'n' type semiconductor is given as,

$$R_H = \frac{-3\pi}{8}\left[\frac{1}{n_e e}\right]$$

$$R_H = \frac{-1.18}{n_e e} \quad ...(3)$$

Conductivity for n type semiconductor is given as,

$$\sigma = n\, e\, \mu_e$$

$$\mu_e = \frac{\sigma_e}{n_e e} \quad ...(4)$$

The above equation (3) can be rewritten as,

$$\frac{1}{n_e e} = \frac{-R_H}{1.18} \quad ...(5)$$

Substituting equation (5) in (4), we get,

$$\mu_e = \frac{-\sigma_e R_H}{1.18} \quad ...(6)$$

Mobility of electron in an n-type semiconductor is given as,

$$\mu_e = -\frac{\sigma_e V_H b}{1.18 \times B} \qquad \left[\because R_H = \frac{V_H b}{I \times B}\right]$$

For p-type semiconductor, the mobility of hole is given as,

$$\mu_e = -\frac{\sigma_h V_H b}{1.18(I \times B)} \quad ...(7)$$

By finding hall voltage, hall coefficient can be calculated and the mobility of the charge carriers is determined.

Application of Hall Effect

- The sign (n-type (or) p-type) of charge carriers can be determined.
- The carrier concentration can be determined as,

$$\left[n = \frac{1.18}{qR_H}\right]$$

- The mobility of charge carriers is measured directly as,

$$\left[\mu = \frac{-\sigma R_H}{1.18}\right]$$

Electrical conductivity can be determined as,

$$[\sigma = n\, q\, \mu]$$

- It is used to determine whether the given material is metal, insulator or semi-conductor and the type of the semiconductor.
- It is used to determine the power flow in an electromagnetic wave.

1.3.2 Generation and Recombination of Charges, Diffusion and Continuity Equation

These are made from a single element. They are also known as indirect band gap semiconductors in which the recombination of free electron from the conduction band with the hole in the valence band takes place via traps.

During recombination, phonons are produced and they heat the crystal lattice (position of the atom). These are the IV group element in the periodic table.

Diffusion

Diffusion is the movement of particles from regions of high concentration to regions of low concentration. To illustrate diffusion in a semiconductor, we consider that the concentration gradient exists in one dimension where the electron density increases with spatial direction x as shown in the below figure:

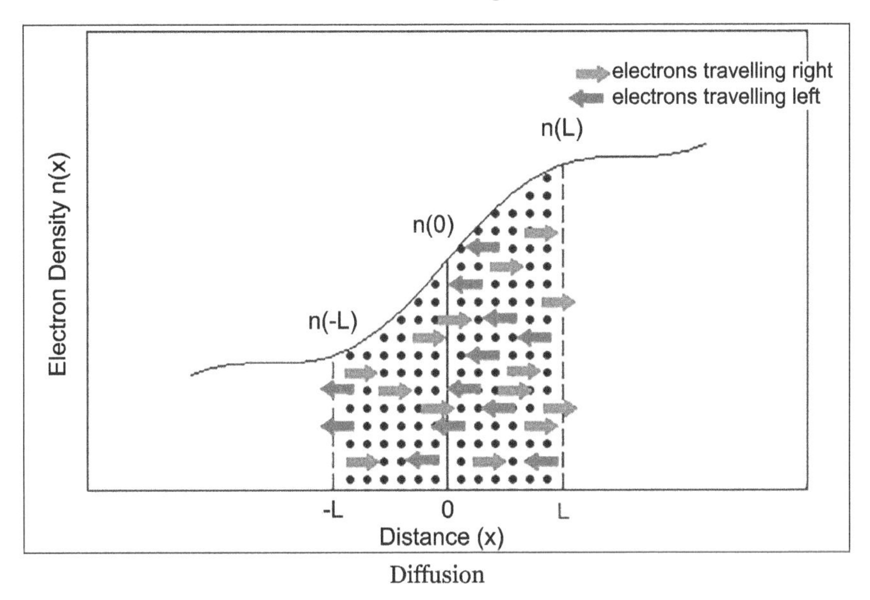

Diffusion

As the electrons move through the semiconductor, they are subjected to many random collisions which change their direction of travel. This gives rise to the concept of a mean free path or mean free time which is the average distance or time taken by an electron to move freely without collision. Since there is no electric field, the electrons travel in any direction with equal probability in between collisions.

In an electron concentration gradient, electrons disperse through the material by random thermal motion. The regions -L and L are one mean free path distance away from the point $x = 0$. Let us determine the net flux of electrons passing through $x = 0$. In the region from $x = (-L, 0)$, half of the electrons will pass across the plane at $x = 0$.

Similarly, half of the electrons in the region from $x = (0, L)$ will pass across the plane at x = 0. If the concentration of electrons in each region are same, then there will be no net flux of electrons across the plane. Since the electron density in each region is different, there is a net flow of electrons across the plane at x = 0.

The flux of electrons in the region -L to 0 is given as,

$$F_1 = \frac{\frac{1}{2} n(-L).L}{\tau_c} = \frac{1}{2} n(-L) v_{th} \quad ...(1)$$

Where,

v_{th} = Thermal velocity of the electrons.

The number of electrons passing through the plane at x=0 from the region 0 to L is given as,

$$F_2 = \frac{1}{2} n(L).v_{th} \quad ...(2)$$

Hence, the net flow of carriers (F) from left to right is (F_1-F_2) which is given as,

$$F = F_1 - F_2 = \frac{1}{2} v_{th} \left[n(-L) - n(L) \right] \quad ...(3)$$

Using a Taylor series expansion with the densities at x = ± L, we have,

$$F = \frac{1}{2} v_{th} \left\{ \left[n(0) - L\frac{dn}{dx} \right] - \left[n(0) + L\frac{dn}{dx} \right] \right\}$$

$$= -v_{th} \frac{dn}{dx} \equiv -D_n \frac{dn}{dx} \quad ...(4)$$

Where,

D_n = Diffusivity

Current associated with diffusion is given as,

$$J_n = -qD_n \frac{dn}{dx} \quad ...(5)$$

In three dimensions, we have,

$$J = -qD_n \left(\frac{\partial n}{\partial x}\hat{i} + \frac{\partial n}{\partial y}\hat{j} + \frac{\partial n}{\partial z}\hat{k} \right) = -qD_n \nabla n \quad ...(6)$$

Continuity Equation

Continuity equation describes the distribution of electrons and holes when there is excess carrier generation, recombination and carrier movement.

As per the law of conservation of charge, rate of change of number of electrons inside the semiconductor is equal to the number of electrons entering per second minus number of electrons leaving per second plus number of electrons generated per second by generation process minus number of electrons lost per second by recombination process.

Derivation of Continuity Equation

Let us consider carrier-flux into/out of an infinitesimal volume.

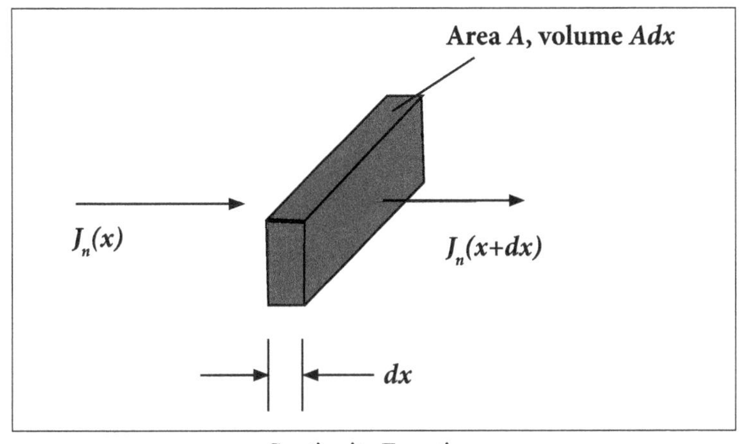

Continuity Equation.

$$Adx\left(\frac{\partial n}{\partial t}\right)=-\frac{1}{q}\left[J_n(x)A-J_n(x+dx)A\right]+G_nAdx-R_nAdx$$

$$J_n(x+dx)=J_n(x)+\frac{\partial J_n(x)}{\partial x}dx$$

$$\Rightarrow\frac{\partial n}{\partial t}=\frac{1}{q}\frac{\partial J_n(x)}{\partial x}-R_n$$

Continuity Equation

$$\frac{\partial n}{\partial t}=\frac{1}{q}\frac{\partial J_n(x)}{\partial x}-R_n+G_L$$

$$\frac{\partial P}{\partial t}=-\frac{1}{q}\frac{\partial J_p(x)}{\partial x}-R_p+G_L$$

The continuity equations based on conservation of carriers is given as,

$$\frac{\partial n}{\partial t} = \frac{1}{q}\frac{\partial J_n(x)}{\partial x} - R_n + G_L$$

$$\frac{\partial p}{\partial t} = \frac{1}{q}\frac{\partial J_n(x)}{\partial x} - R_p + G_L$$

The minority carrier diffusion equations are derived from the continuity equations. For minority carriers under certain conditions (small E-field, low-level injection, uniform doping profile), we have,

$$\frac{\partial \Delta n_p}{\partial t} = D_N \frac{\partial^2 \Delta n_p}{\partial x^2} - R_n + G_L \quad \frac{\partial \Delta p_n}{\partial t} = D_P \frac{\partial^2 \Delta p_n}{\partial x^2} - R_p - G_L$$

1.3.3 Injected Minority Carriers

Minority carrier injection in electronics is a process that takes place at the boundary between p-type and n-type semiconductor materials which is used in some types of transistors. Each semiconductor material contains two types of freely moving charges namely electrons and holes.

Electrons are more abundant or majority carriers and holes are less abundant or minority carrier in n-type materials. In p-type materials, holes are the majority carriers and electrons are the minority carrier.

If a battery is properly connected to the semiconductor material, the p-type material acquires additional electrons which is injected into the p-type material from the n-type material by the flow of the electrons from the battery. This is called as minority carrier injection which is important in bipolar junction transistors that are made up of two p-n junctions.

Consider a long semiconductor bar which is doped uniformly with donor atoms, as shown in the below figure:

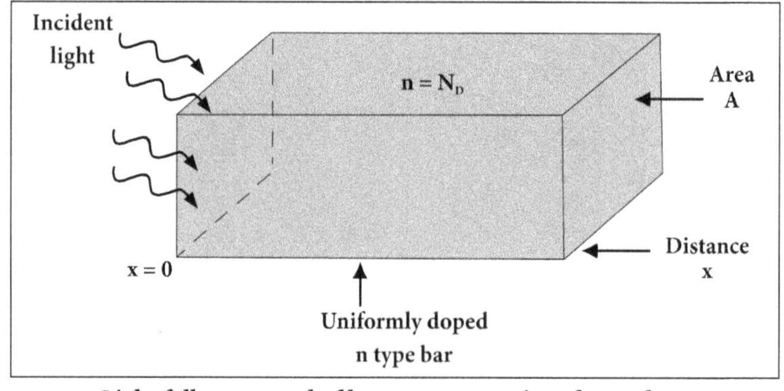

Light falls upon end of long n-type semiconductor bar.

The concentration $n \approx N_D$ and is independent of position as bar is uniformly doped.

The end on which the light is incident is denoted as x = o.

Due to photoexcitation, the covalent bonds are broken at the surface, which is radiated. Due to this, the electron-hole pairs are generated and are injected into the bar at x = o.

The bar is n-type and free electrons are large in number. Thus generated electron-hole pairs mainly affect the minority charge concentration i.e. concentration of holes. Let us analyze the behaviour of steady state minority carrier concentration p against the distance x in the bar.

Let p' = concentration of injected minority charge carriers.

Assumption: It is assumed that the concentration p' is much smaller than the doping level i.e. p' <<n. This condition stating that minority concentration is much smaller than majority concentration and is called low level injection conduction.

It is known that $J=(n\mu_n+p\mu_p)q\,E$, thus the drift current can be neglected compared to electron drift current.

Now $p=p'+p_o \ll n$ hence the hole drift current can be neglected compared to electron drift current.

It can be assumed that the hole current Ip is entirely because of diffusion. According yo continuity equation, the controlling differential equation for minority carrier concentration p is,

$$\frac{d^2p}{dx^2}=\frac{p-p_o}{D_p\tau_p} \qquad ...(1)$$

The diffusion length for the holes is Lp and given by,

$$L_p=\sqrt{D_p\tau_p} \qquad ...(2)$$

Hence the differential equation for the injected hole concentration $p'=p-p_o$ is given by,

$$\frac{d^2p'}{dx^2}=\frac{p'}{L_p^2} \qquad ...(3)$$

Solution of this differential equation is given by,

$$P'(x)=K_1\,e^{-x/L_P}+K_2\,e^{+x/L_P} \qquad ...(4)$$

where K_1, K_2 = Constants of integration.

Now as $x \to \infty$, the concentration cannot become infinite. Thus the second term must be zero in the equation (4), for which $K_2 = 0$.

At x = 0, the injected concentration is p'(0). Hence using in the equation (4),

$$p'(0) = K_1\ e^{0} \text{ i.e. } K_1 = p'(0),$$

$$p'(x) = p'(0)\ e^{-x/L_P} = P(x) - P_0 \qquad ...(5)$$

$$p(x) = p_0 + p'(0)e^{-x/L_P} \qquad ...(6)$$

The equation (6) represents that the hole concentration decreases exponentially with the distance. This is shown in the below figure:

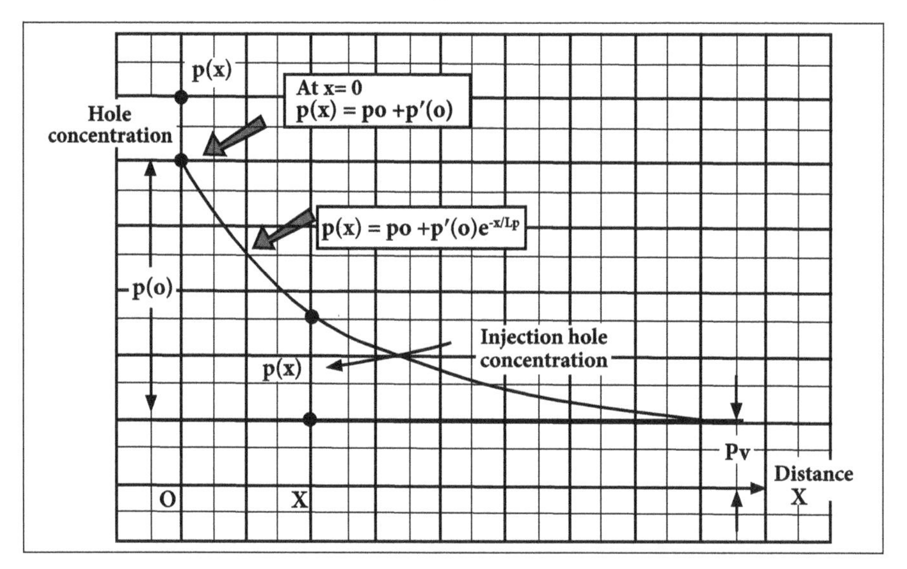

Exponential behaviour of hole concentration in a n-type semiconductor bar.

1.3.4 Law of Junction

$$n_p(-x_p) = N_d e^{\left[\frac{-\phi B}{V_{th}}\right]} e^{\left[\frac{V_D}{V_{th}}\right]} = n_{po} e^{\left[\frac{V_D}{V_{th}}\right]}$$

$$P_n(x_n) = N_a e^{\left[\frac{-\phi_B}{V_{th}}\right]} e^{\left[\frac{V_D}{V_{th}}\right]} = p_{no} e^{\left[\frac{V_D}{V_{th}}\right]}$$

Where,

$$n_{po} = \frac{n_i^2}{N_a} \text{ and } P_{no} = \frac{n_i^2}{N_d}$$

Minority carrier concentration at the SCR is an exponential function of applied bias. It changes one decade for every 60mV change in V_D.

- Law of the junction is valid if minority carrier concentration is less than the majority carrier concentration. This condition is called as low level injection.

$$p_n < n_{no} \quad \text{and} \quad n_p < p_{po}$$

1.4 Introduction to Fermi Level in Intrinsic and Extrinsic Semiconductors with Necessary Mathematics

Fermi level and variation of fermi level with temperature in an intrinsic semiconductor:

Derivation of Fermi Level

In an intrinsic semiconductor,

Density of electrons (N_e) = Density of holes (N_h)

$$2\left[\frac{2\pi m_e^* K_B T}{h^2}\right]^{3/2} \exp\left[\frac{E_F - E_C}{K_B T}\right] = 2\left[\frac{2\pi m^* h K_B T}{h^2}\right]^{3/2} \exp\left[\frac{E_V - E_F}{K_B T}\right]$$

$$\left[m_e^*\right]^{3/2} \exp\left[\frac{E_F - E_C}{K_B T}\right] = \left[m_h^*\right]^{3/2} \exp\left[\frac{E_v - E_F}{K_B T}\right]$$

$$\left[\frac{m_h^*}{m_e^*}\right]^{3/2} = \exp\left[\frac{E_F - E_C - E_V + E_F}{K_B T}\right]$$

$$\left[\frac{m_h^*}{m_e^*}\right]^{3/2} = \exp\left[\frac{2E_F - E_C - E_V}{K_B T}\right]$$

Taking log on both sides, we get,

$$\log\left[\frac{m_h^*}{m_c^*}\right] = \frac{\left[2E_F - E_C - E_V\right]}{K_B T}$$

$$\frac{3}{2}\log\left[\frac{m_h^*}{m_e^*}\right] = \frac{\left[2E_F - E_C - E_V\right]}{K_B T}$$

$$\frac{3}{2} K_B T \log\left[\frac{m_h^*}{m_e^*}\right] = \left[2E_F - (E_C + E_V)\right]$$

$$2E_F = (E_C + E_V) + \frac{3}{2} K_B T \log\left[\frac{m_h^*}{m_e^*}\right]$$

$$E_F = \frac{E_C + E_V}{2} + \frac{3}{4} K_B T \log\left[\frac{m_h^*}{m_c^*}\right]$$

If $m_h^* = m_e^*$, then at $T = o\ K, \frac{3}{2} K_B T \log\left[\frac{m_h^*}{m_e^*}\right]$

$$E_F = \left[\frac{E_C + E_V}{2}\right]$$

Fermi level lies in the midway between the conduction level (E_c) and valence level (E_v) at T = oK.

If $m_h^* = m_e^*$, then the fermi level is a function of temperature and it is raised slightly with the temperature.

1.4.1 Extrinsic Semiconductors with Necessary Mathematics

An extrinsic semiconductor is a semiconductor which is doped by a specific impurity to modify its electrical properties which is suitable for electronic applications or optoelectronic applications.

p-type Semiconductors

A p-type semiconductor is an intrinsic semiconductor (like Si) in which an impurity acts as an acceptor (e.g. Boron (B) in Si). These impurities are called as acceptors because, once they are inserted in the crystalline lattice, they lack one or several electrons to realize a full bonding with the rest of the crystal.

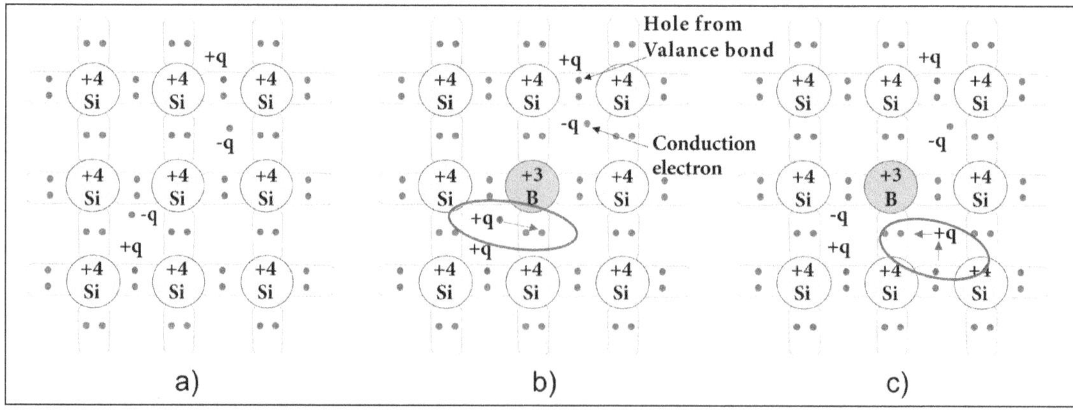

Si crystal doped with boron (B).

From the above figure, a p-type semiconductor has a lower electron density (n) and a higher hole density (p) than the same intrinsic semiconductor. Electrons are called as the minority carriers where holes are known as the majority carriers.

For an extrinsic semiconductor, the dopant density is always higher than the intrinsic carrier density $(N_A >> n_i)$. In a p-type material, the hole density is close to the dopant density N_A. Since the law of mass action is always true, expressions for the carrier densities are given as,

$$n = \frac{n_i^2}{N_A}$$

$$p = N_A$$

Fermi level for a p-type semiconductor or chemical potential is given as,

$$E_{Fp} = E_v + kT \ln \frac{N_v}{N_A}$$

When the acceptor density is increased, fermi level moves closer to the edge of the valence band. If $N_A = N_v$, the fermi level enters valence band and the semiconductor is said to be degenerated.

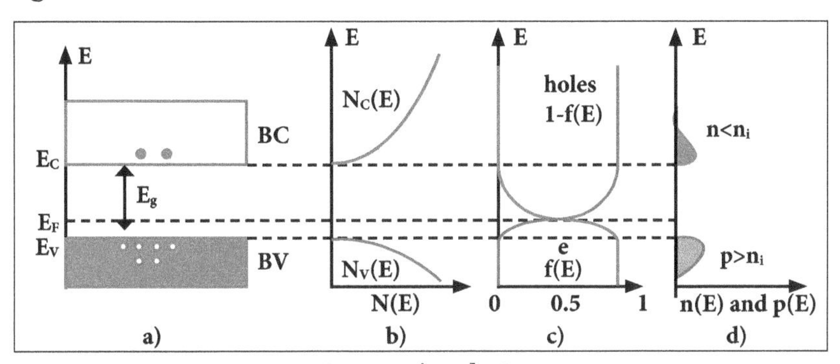

p-type semiconductor.

n-type Semiconductors

n-type semiconductor is an intrinsic semiconductor (e.g. silicon Si) in which a donor impurity is present. The impurities are known as donor impurities since they give an extra electron to the conduction band to make bonds with the neighboring atoms.

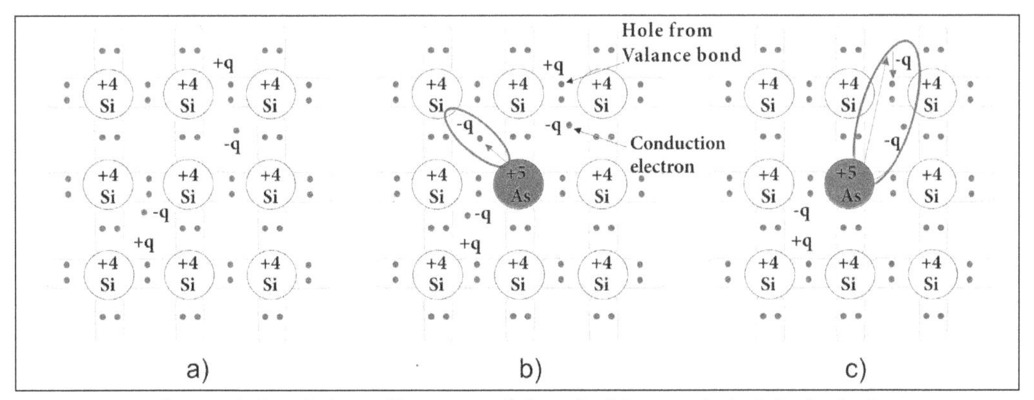

Electronic bonds in a Silicon crystal doped with Arsenic (As) (n doping).

n-type semiconductor has a higher electron density (n) and a lower hole density (p) than the same intrinsic semiconductor. Holes are known as the minority carriers whereas electrons are called as the majority carriers.

In p-type semiconductors,

$$n = N_D$$

$$p = \frac{n_i^2}{N_D}$$

Where,

$$N_D = \text{Donor density}$$

Fermi level of a n-type semiconductor is given as,

$$E_{Fn} = E_c - kT \ \text{In} \ \frac{N_c}{N_D}$$

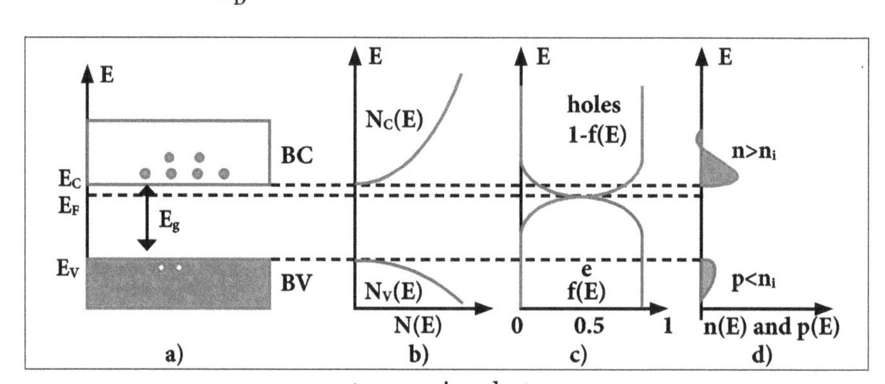

n-type semiconductor.

When the donor density is increased, the fermi level moves closer to the edge of the conduction band. If $N_D = N_c$, then the fermi level enters into the conduction band and the semiconductor is said to be degenerated.

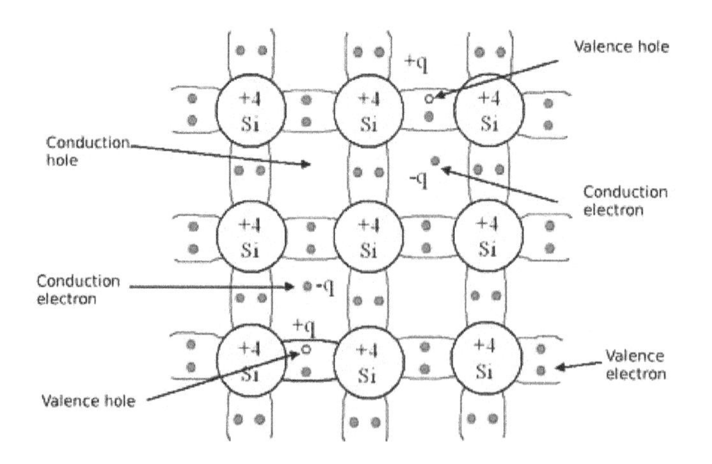

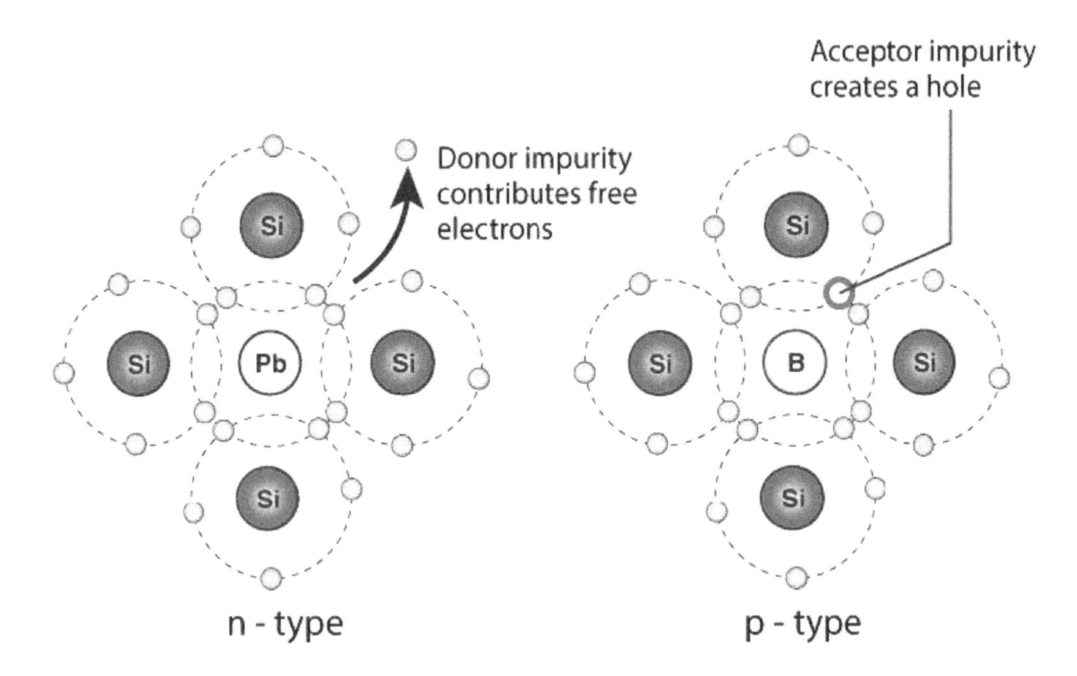

Extrinsic Semiconductors

2

Junction Diode Characteristics

2.1 Operation and Characteristics of p-n Junction Diode

Structure

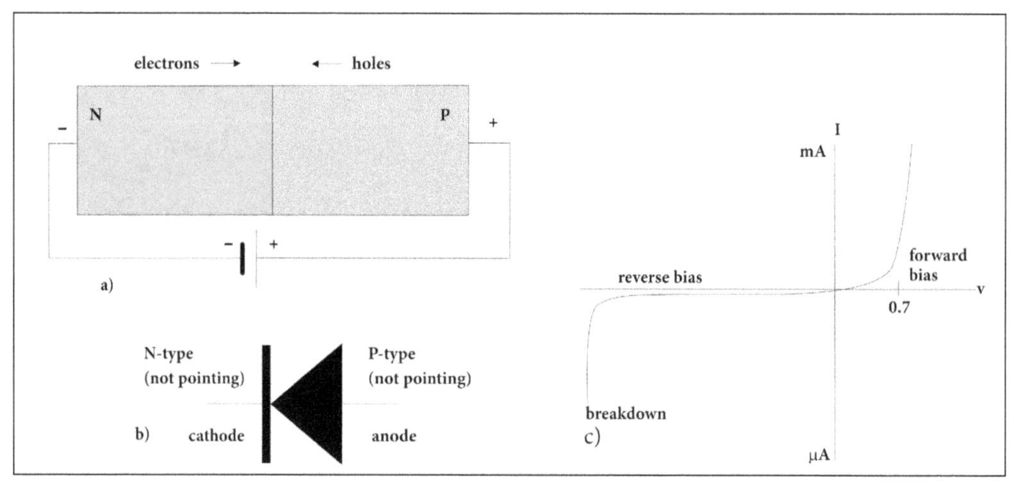

Symbol of PN junction diode.

- The p region has a holes (Majority carriers).
- The n region has a free electrons.
- The p and n charge carriers are called as Mobile charges.

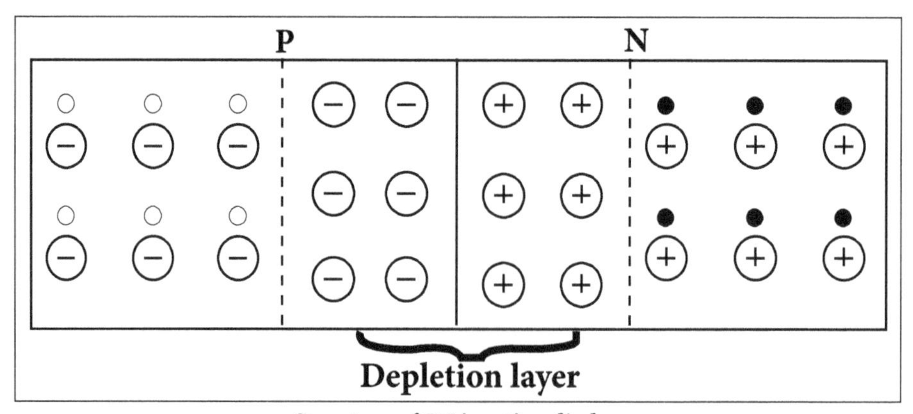

Structure of PN junction diode.

PN Junction is formed when some of the holes in P-region and the free electrons in the N-region diffuse into each other and disappear due to recombination.

Knee Voltage of Diode

The voltage above which the diode operates in normal region is called as Knee voltage of a diode.

PN Junction Diode Act as Rectifier

A PN Junction diode is a two terminal device that is polarity sensitivity. When the diode is forward biased, the diode conducts and allows current to flow through it without any resistance. i.e., the diode is ON.

When the diode is reverse biased, the diode does not conduct and no current flows through it i.e., the diode is OFF or providing the blocking function. Thus, an ideal diode acts as a switch; either opened or closed depending upon the polarity of the voltage placed across it.

The ideal diode has zero resistance under forward bias and infinite resistance under reverse bias.

Avalanche Breakdown in Transistor

When electrons and holes in semiconductors are accelerated by applied voltage, these accelerated holes and electron collide with the bound electrons and produces free electrons. This causes even more collision due to increase in current value. This is called as Avalanche breakdown.

Transition Capacitance of the Diode

A reverse biased condition capacitance exists at the PN junction, where the two regions act as the plates while the depletion region acts as dielectric is called as transition capacitance of a diode.

Reverse Saturation Current

The current due to the minority carriers in reverse bias is said to be reverse saturation current. This current depends on the value of the reverse bias voltage.

2.1.1 Current Components in p-n Diode and Diode Equation

Current Components in p-n Diode

It is indicated earlier that when a p-n junction diode is forward biased, large forward current flows, which is mainly due to the majority carriers. The depletion region near the junction is extremely small, under forward biased biased condition. In forward biased condition holes get diffused into n-side from p-side and electrons get diffused into p-side from n-side. So on p-side, the current carried by electrons called the diffusion current due to minority carriers, gets decreases exponentially with respect to the distance measured from the junction. This current due to electrons, on p-side which are minority carriers is denoted as I_{np} Same way the holes from p-side diffuse into n-side carry current which decreases exponentially with respect to distance measured from the junction. The current due to holes on n-side, which are minority carriers is denoted as I_{pn} If distance is denoted by x then,

- $I_{np}(x) =$ Current due to electrons in p-side as a function of x.
- $I_{pn}(x) =$ Current due to holes in n-side as a function of x.

At the junction i.e. at x = 0, electrons crossing from n-side to p-side constitute a current, $I_{np}(0)$ in the same direction as holes crossing the junction from p-side to n-side constitute a current, $I_{pn}(0)$.

Hence the current at the junction is the total conventional current I flowing through the circuit.

$$I = I_{pn}(0) + I_{np}(0) \qquad ...(1)$$

Now $I_{pp}(x)$ decreases on n-side as we move away from junction on n-side. Similarly $I_{np}(x)$ decreases on p-side as we move away from junction on p-side.

But as the entire circuit is a series circuit, the total current must be maintained at, independent of x. This indicates that on p-side there exists one more current component which is due to holes on p-side which are the majority carriers. It is denoted by $I_{pp}(x)$ and the addition of the two currents on p-side is total current I.

$I_{pp}(x) =$ Current due to holes in p-side.

Similarly on n-side, there exists one more current component which is due to electrons on n-side, which are the majority carriers. It is denoted as $I_{nn}(x)$ and the addition of the two currents on n-side is total current I.

$I_{nn}(x) =$ Current due to electrons in n-side.

$$\text{On p-side, } I = I_{pp}(x) + I_{np}(x) \qquad ...(2)$$

$$\text{On n-side, } I = I_{nn}(x) + I_{pn}(x) \qquad ...(3)$$

These current components are plotted as a function of distance in the figure below:

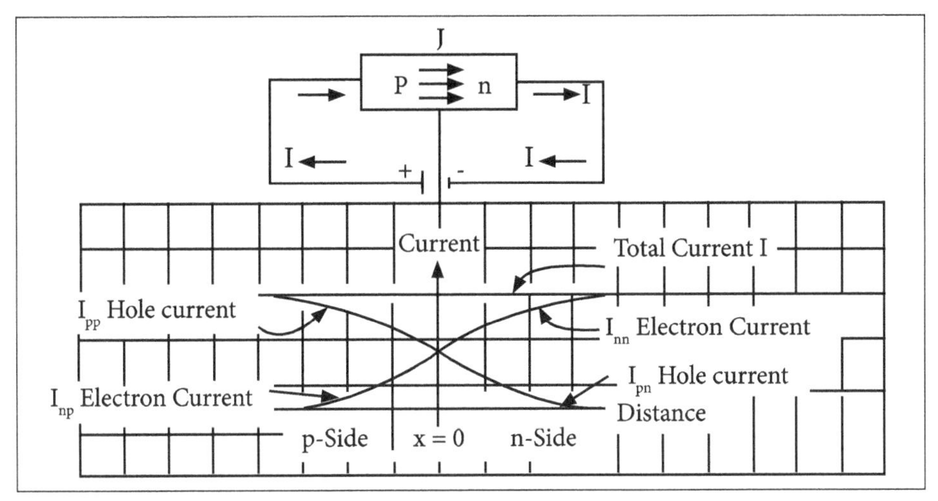

Current components.

The current I_{pp} decreases towards the junction, at the junction enters the n-side and becomes I_{pn} which further decreases exponentially. Similarly the current Inn decreases towards the junction, at the junction enters the p-side and becomes I_{np} which also again decreases exponentially.

Note: In forward bias condition, current enters the p-side as a hole current. So sum of the currents carried by electrons and holes at any point inside the diode is always constant equal to the total forward current I. But the proportion due to holes and the electrons in constituting the current varies with the distance, from the junction.

Diode Equation

The ideal diode equation is an equation that represents the current flow through an ideal p-n junction diode as a function of applied voltage. In realistic settings, current will deviate slightly from this ideal case.

Ideal Diode Equation

A p-n junction diode creates the following current such as: under reverse bias, there is a small, constant reverse current, and under forward bias, there is a forward current that increases with voltage. The current-voltage function for an ideal diode is,

$$i(v) = I_S\left[\exp(v/\eta V_T) - 1\right],\ v > VZ$$

Where I_s is the reverse saturation current, v is the applied voltage, $V_T = T/11{,}586$ is the volt equivalent of temperature, and η is the emission coefficient, which is 1 for germanium devices and 2 for silicon devices. Note that i is defined as positive when flowing from p to n. This is also called the Shockley ideal diode equation or the diode law. Note also that for $v \leq V_Z$, the diode is in breakdown and the ideal diode equation no longer applies; for $v \leq V_Z$, $i = -\infty$. The ideal diode i-v characteristic curve is shown below:

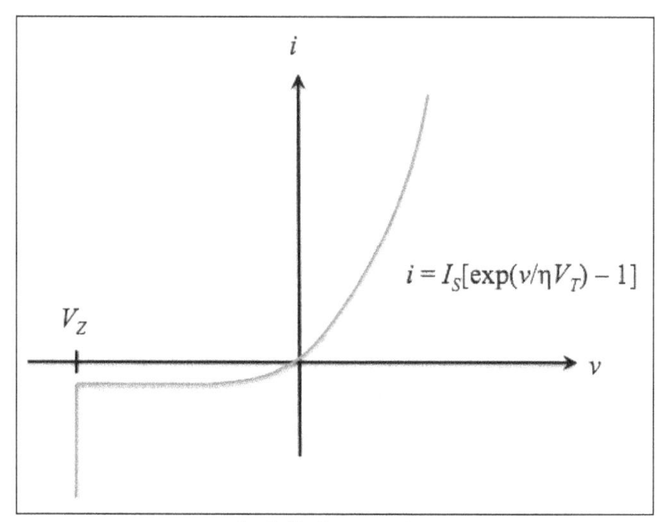

Ideal diode equation.

The ideal diode equation is very useful as a formula for current as a function of voltage. However, at times the inverse relation may be more useful; if the ideal diode equation is inverted and solved for voltage as a function of current, we find:

$$v(i) = \eta V_T \ln\left[(i/I_S)+1\right].$$

2.1.2 Temperature Dependence on V-I Characteristics

The general expression for the electrical conductivity, $\sigma = n\, e\, \mu$

The intrinsic electrical conductivity, $\sigma_i = [n\, e\, \mu_e + p\, e\, \mu_h]$

But,

$$n = p = n_i$$

$$\sigma_i = [n_i\, e\mu_e + n_i\, e\mu_h]$$

$$\sigma_i = n_i\, e[\mu_e + \mu_h]$$

Where,

μ_e – Electron mobility

μ_h – Hole mobility

$$\sigma_i = [\mu_e + \mu_h] e \; 2\left[\frac{2\pi K_B T}{h^2}\right]^{3/2} \left[m_e^* m_h^*\right]^{3/4} \exp\left[-\frac{E_g}{2K_B T}\right]$$

The electrical conductivity depends on the negative exponential of band gap Eg between the valance band and conduction band and also for the mobilities of both holes and electrons. The mobilities in a pure semiconductor are determined by the interaction of electron with lattice waves or phonons.

So that we can neglect $[\mu_e + \mu_h]$.

The electrical conductivity is,

$$\sigma_i = C \exp\left[-\frac{E_g}{2K_B T}\right] \qquad ...(1)$$

Where C is a constant.

Taking log on both sides of equation (1),

$$\log \sigma_i = \text{Log}\left[C \exp\left(-\frac{E_g}{2K_B T}\right)\right]$$

$$\log \sigma_i = \text{Log } C + \left(\frac{-E_g}{2K_B T}\right)$$

$$\log \sigma_i = \text{Log } C - \left(\frac{E_g}{2K_B T}\right)$$

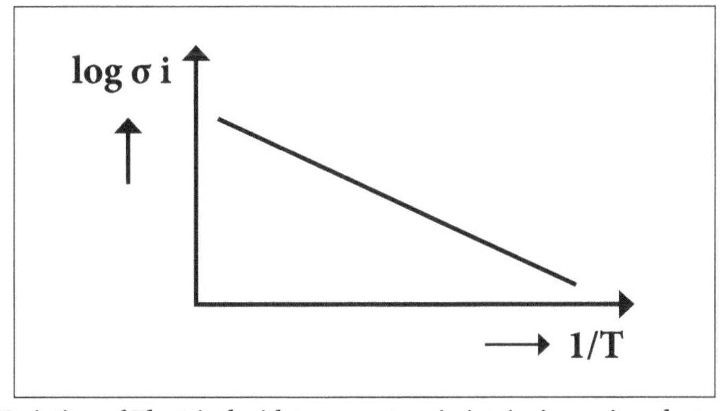

Variation of Electrical with temperature in intrinsic semiconductor.

A graph is drawn between 1/T and Log σ_i. From the graph, it is noted that this electrical conductivity increases with temperature.

2.2 Diffusion Capacitance and Diode Resistance

When a PN junction is formed, a layer of positive and negative impurity ions called as depletion layer is formed on either side of the junction. It is also known as depletion-region, space-charge region or transition region.

The depletion-layer acts as a dielectric medium between P-region and N-region. These regions acts as two plates of a capacitor separated by a dielectric.

The capacitance formed in a junction area is called as depletion layer capacitance. It is also called as depletion region-capacitance, space charge capacitance, transition region capacitance or simply junction capacitance.

The capacitance of a parallel plate capacitor a given by the equation,

$$C = \frac{\varepsilon \cdot A}{d}$$

Where e is the permittivity of the dielectric between the plates of area A separated by a distance d.

Since the depletion layer width (d) increases with the increase in reverse bias voltage, the resulting depletion layer capacitance will decreases with the increased reverse bias.

The depletion layer capacitance depends upon the nature of a PN junction, semiconductor material and magnitude of the applied reverse voltage. It is given by the relation,

$$C_T = \frac{k}{(V_B - V)^n}$$

Where,

k = A constant, depending upon the nature of semiconductor

V_B = Barrier Voltage. It is 0.6 V for silicon and 0.2 V for Ge

V = Applied reverse voltage

n = A constant depending upon the nature of the junction

Diffusion Capacitance

The capacitance which exists in a forward-biased junction is called as diffusion or storage capacitance. It is different from the depletion layer capacitance which exists in a reverse-biased junction.

The diffusion capacitance arises due to the arrangement of minority carrier density and its value is much larger than the depletion layer capacitance. Its value for an abrupt junction is given by,

$$C_D = \frac{\tau \cdot I}{\eta \cdot V_T}$$

Where,

$\tau =$ Mean life time of the carriers

$I =$ Value of forward current

$\eta =$ A constant (1 for Ge)

$V_T =$ Volt equivalent of temperature

It is evident from the above relation that diffusion capacitance is directly proportional to the forward current.

2.2.1 Energy Band Diagram of p-n Diode

Carrier Concentration In p -Type Semiconductor

If trivalent (Aluminum, Gallium, Indium) impurities are doped with pure semiconducting material the holes are produced, this is called P - type semiconductor.

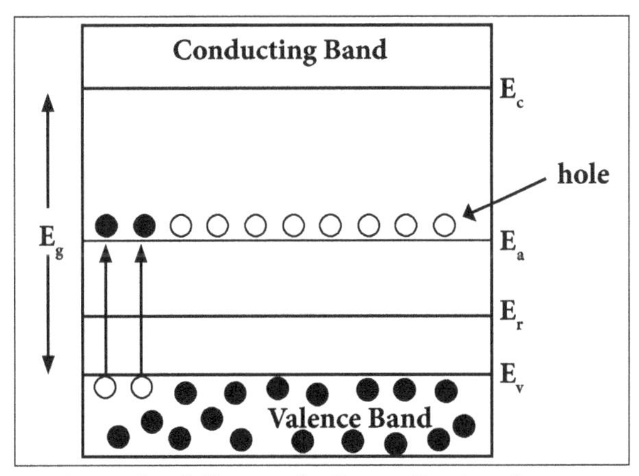

Energy level diagram for P-type Semiconductor.

We know that,

Density of holes in the valence band in an intrinsic semiconductors is,

$$N_h = 2\left[\frac{2\pi m_h^* K_B T}{h^2}\right]^{3/2} \exp\left[\frac{E_v - E_F}{K_B T}\right] \qquad ...(1)$$

Put,

$$N_v = 2\left[\frac{2\pi m_h^* K_B T}{h^2}\right]^{3/2}$$

Density of holes is $N_h = N_v \exp\left[\frac{E_V - E_F}{K_B T}\right]$ ...(3)

Density of ionized acceptor atoms is $N_A\ F[E_A] = N_A\ \exp\left[\frac{E_F - E_A}{K_B T}\right]$ ...(4)

At equilibrium condition,

[Number of holes per unit volume in valence band equation (3)] = [Number of electron per unit volume in acceptor energy level equation (4)].

Taking log on both sides, we get,

$$N_v \exp\left[\frac{E_V - E_F}{K_B T}\right] = N_A \exp\left[\frac{E_F - E_A}{K_B T}\right]$$

$$\exp\left[\frac{E_v - E_F}{K_B T}\right] \exp\left[\frac{-[E_F - E_A]}{K_B T}\right] = \frac{N_A}{N_v}$$

$$\exp\left[\frac{E_v - E_F - E_F + E_A}{K_B T}\right] = \frac{N_A}{N_v}$$

Substituting the value N_v,

$$\log\left[\exp\left[\frac{E_v - E_F - E_F + E_A}{K_B T}\right]\right] = \log\left[\frac{N_A}{N_v}\right]$$

$$\left[\frac{E_v - E_F - E_F + E_A}{K_B T}\right] = \log\left[\frac{N_A}{N_v}\right]$$

$$E_v - E_F - E_F + E_A = K_B T \log\left[\frac{N_A}{N_v}\right]$$

$$-2E_F = -[E_V + E_A] + K_B T \log\left[\frac{N_A}{N_v}\right]$$

$$E_F = \frac{(E_V + E_A)}{2} - \frac{K_B T}{2} \log\left[\frac{N_A}{N_v}\right]$$

At T = 0 K,

$$E_F = \frac{(E_v + E_A)}{2} - \frac{K_B T}{2} \log \left[\frac{N_A}{2\left(\frac{2\pi m_h^* K_B T}{h^2}\right)^{3/2}} \right] \qquad ...(5)$$

At 0 K fermi level in p-type semiconductor lies exactly at the middle of the acceptor level and the top of the valance band.

$$E_F = \left[\frac{(E_v + E_A)}{2}\right]$$

Expression for the Density of Holes in Valence Band in Terms of N_A

As the temperature is increased more and more, the acceptor atoms are ionized. Further increase in temperature results in generation of electron hole pairs due to breaking of covalent bonds and materials tends to behave in an intrinsic manner. The fermi level gradually moves towards the intrinsic fermi level.

We know density of holes in valence band is,

$$N_h = 2\left[\frac{(2\pi m_h^* K_B T)}{h^2}\right]^{3/2} \exp\left[\left(\frac{E_v - E_F}{K_B T}\right)\right] \qquad ...(1)$$

Substituting the equation (5) in (1) we get,

$$N_h = 2\left(\frac{2\pi m_h^* K_B T}{h^2}\right) \exp - \frac{K_B T}{2} \log \left[\frac{E_V - \left[\left(\frac{E_v + E_A}{2}\right) - \frac{K_B T}{2} \log \frac{N_A}{2\left(\frac{2\pi m_h^* K_B T}{h^2}\right)^{3/2}}\right]}{K_B T} \right]$$

$$= 2\left[\frac{(2\pi m_h^* K_B T)}{h^2}\right]^{3/2} \exp\left[\left(\frac{E_v - E_A}{2K_B T}\right) + \frac{1}{2}\log\left(\frac{N_A}{2\left[\frac{[2\pi m_h^* K_B T]}{h^2}\right]^{3/2}}\right)\right]$$

$$= 2\left[\frac{\left(2\pi m_h^* K_B T\right)}{h^2}\right]^{3/2} \exp\left[\left(\frac{E_v - E_A}{2K_B T}\right) + \log\left(\frac{N_A}{2\left[\frac{\left[2\pi m_h^* K_B T\right]}{h^2}\right]^{3/2}}\right)^{1/2}\right]$$

$$= 2\left[\frac{\left(2\pi m_h^* K_B T\right)}{h^2}\right]^{3/2} \exp\left[\left(\frac{E_v - E_A}{2K_B T}\right) + \log\left(\frac{N_A^{1/2}}{2^{1/2}\left[\frac{\left[2\pi m_h^* K_B T\right]}{h^2}\right]^{3/2}}\right)\right]$$

$$= 2\left[\frac{\left(2\pi m_h^* K_B T\right)}{h^2}\right]^{3/2} \exp\left(\frac{E_v - E_A}{2\, K_B T}\right)\left(\frac{N_A^{1/2}}{2^{1/2}\left[\frac{\left[2\pi m_h^* K_B T\right]}{h^2}\right]^{3/4}}\right)$$

$$= \left(2N_A\right)^{1/2} \exp\left[\frac{E_v - E_A}{2\, K_B T}\right]\left[\frac{\left(2\pi m_h^* K_B T\right)}{h^2}\right]^{3/2}\left[\frac{\left(2\pi m_h^* K_B T\right)}{h^2}\right]^{-3/4}$$

$$= \left(2N_A\right)^{1/2} \exp\left[\frac{E_v - E_A}{2K_B T}\right]\left[\frac{\left(2\pi m_h^* K_B T\right)}{h^2}\right]^{3/4}$$

Here $E_A - E_V = \Delta E$ is known as ionization energy of acceptors.

i.e. ΔE represents the energy required for an electron to move from valance band (E_V) to acceptor energy level (E_A).

$$N_h = \left(2N_A\right)^{1/2} \exp\left[\frac{-\Delta E}{2K_B T}\right]\left[\frac{\left(2\pi m_h^* K_B T\right)}{h^2}\right]^{3/4}$$

Carrier Concentration In n-Type Semiconductor

If pentavalent (Phosphorous, Arsenic, Antimony) impurities are doped with pure semiconducting material the free electrons are produced, this is called N-type semiconductor.

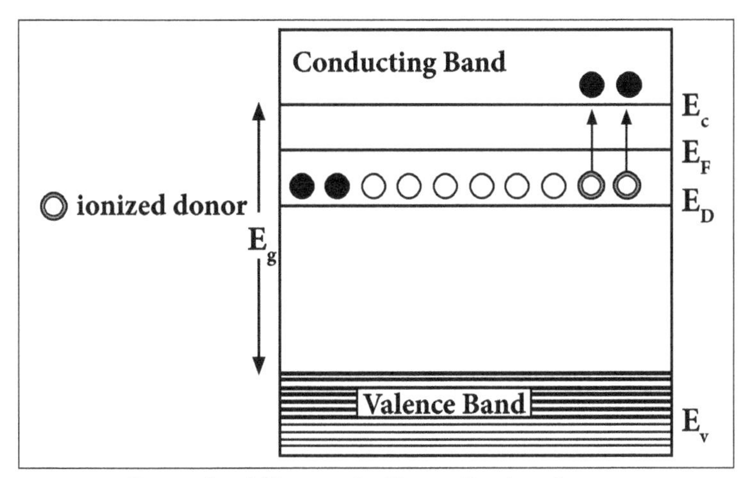

Energy level diagram for N-type Semiconductor.

We know that,

Density of electrons in conduction band in an intrinsic semiconductor is,

$$N_e = \left[2\left[\frac{2\pi m_e^* K_B T}{h^2}\right]^{3/2} \exp\left[\frac{E_F - E_c}{K_B T}\right]\right] \qquad ...(1)$$

Put,

$$N_c = 2\left[\frac{2\pi m_e^* K_B T}{h^2}\right]^{3/2}$$

Density of electrons is,

$$N_e = N_c \exp\left[\frac{E_F - E_c}{K_B T}\right] \qquad ...(2)$$

Density of ionized donor atoms is given by,

$$N_D\left[1 - F(E_D)\right] = N_D \exp\left[\frac{E_D - E_F}{K_B T}\right] \qquad ...(3)$$

At equilibrium condition,

[Number of electrons per unit volume in conduction band equation (2)] = [Number of holes per unit volume in donor energy level equation (3)]

$$N_c \exp\left[\frac{E_F - E_C}{K_B T}\right] = N_D \exp\left[\frac{E_D - E_F}{K_B T}\right]$$

$$\frac{\exp\left[\dfrac{E_F - E_c}{K_B T}\right]}{\exp\left[\dfrac{E_D - E_F}{K_B T}\right]} = \frac{N_D}{N_c}$$

$$\exp\left[\frac{E_F - E_c - E_D + E_F}{K_B T}\right] = \frac{N_D}{N_c}$$

Taking log on both sides, we get,

$$\log\left[\exp\left[\frac{E_F - E_c - E_D + E_F}{K_B T}\right]\right] = \log\left[\frac{N_D}{N_c}\right]$$

$$\left[\frac{E_F - E_c - E_D + E_F}{K_B T}\right] = \log\left[\frac{N_D}{N_c}\right]$$

$$2E_F = (E_c + E_D) + K_B T \log\left[\frac{N_D}{N_c}\right]$$

$$E_F = \frac{(E_c + E_D)}{2} + \frac{K_B T}{2}\log\left[\frac{N_D}{N_c}\right] \qquad ...(4)$$

At T = 0 K,

$$E_F = \left[\frac{(E_C + E_D)}{2}\right] \qquad ...(5)$$

At T = 0 K. Thus, the Fermi level in N-type semiconductor lies exactly in middle of the conduction level (E_C) and donor level (E_D).

This equation shows that the electron concentration in the conduction band is proportional to the square root of the donor concentration.

Expression for the Density of Electrons in Conduction Band in Terms of N_D

As the temperature is increased more and more, the donor atoms are ionized and the fermi level drops. For a particular temperature all donor atoms are ionized, further increase in temperature results in generation of electron hole pairs due to breaking of covalent bonds and materials tends to behave in an intrinsic manner.

We know density of electrons in conduction band is,

$$N_e = 2\left[\frac{(2\pi m_e^* K_B T)}{h^2}\right]^{3/2} \exp\left[\frac{E_F - E_C}{K_B T}\right] \qquad ...(1)$$

Substituting the equation (4) in (1) we get,

$$N_e = 2\left[\frac{(2\pi m_e^* K_B T)}{h^2}\right]^{3/2} \exp\left[\frac{\left(\frac{E_c + E_D}{2}\right) + \frac{K_B T}{2}\log\left(\frac{N_D}{2\left[\frac{[2\pi m_e^* K_B T]}{h^2}\right]^{3/2}}\right) - E_c}{K_B T}\right]$$

$$= 2\left[\frac{(2\pi m_e^* K_B T)}{h^2}\right]^{3/2} \exp\left[\left(\frac{E_C + E_D}{2}\right) + \frac{1}{2}\log\left(\frac{N_D}{2\left[\frac{[2\pi m_e^* K_B T]}{h^2}\right]^{3/2}}\right) - \frac{E_c}{K_B T}\right]$$

$$= 2\left[\frac{(2\pi m_e^* K_B T)}{h^2}\right]^{3/2} \exp\left[\left(\frac{E_D - E_C}{2 K_B T}\right) + \log\left(\frac{N_D}{2\left[\frac{[2\pi m_e^* K_B T]}{h^2}\right]^{3/2}}\right)^{1/2}\right]$$

$$= 2\left[\frac{(2\pi m_e^* K_B T)}{h^2}\right]^{3/2} \exp\left[\left(\frac{E_D - E_C}{2 K_B T}\right) + \log\left(\frac{N_D^{1/2}}{2^{1/2}\left[\frac{[2\pi m_e^* K_B T]}{h^2}\right]^{3/4}}\right)\right]$$

$$= 2\left[\frac{(2\pi m_e^* K_B T)}{h^2}\right]^{3/2} \exp\left(\frac{E_D - E_C}{2 K_B T}\right)\left(\frac{N_D^{1/2}}{2^{1/2}\left[\frac{[2\pi m_e^* K_B T]}{h^2}\right]^{3/4}}\right)$$

$$= (2N_D)^{1/2} \exp\left[\frac{E_D - E_C}{2\, K_B T}\right]\left[\frac{(2\pi m_e^* K_B T)}{h^2}\right]^{3/2}\left[\frac{(2\pi m_e^* K_B T)}{h^2}\right]^{-3/4}$$

$$= (2N_D)^{1/2} \exp\left[\frac{E_D - E_C}{2K_B T}\right]\left[\frac{(2\pi m_e^* K_B T)}{h^2}\right]^{3/4}$$

Here $E_C - E_D = \Delta E$ is known as ionization energy of donors i.e. ΔE represents the amount of energy required to transfer on an electron to from donor energy level (E_D) to conduction band (E_C).

$$N_e = (2N_D)^{1/2} \exp\left[\frac{-\Delta E}{2\, K_B T}\right]\left[\frac{(2\pi m_e^* K_B T)}{h^2}\right]^{3/4}$$

2.3 Special Diodes: Avalanche and Zener Breakdown

Avalanche Breakdown

When electrons and holes in semiconductors are accelerated by applied voltage, these accelerated holes and electron collide with the bound electrons and produces free electrons. This causes even more collision due to increase in current value. This is called as Avalanche breakdown.

Zener Breakdown

The Zener Breakdown is observed in Zener diodes having V_z less than 5V or between 5 to 8 volts. When a reverse voltage is applied to a Zener diode, it causes very intense electric field to appear across a narrow depletion region. Such intense electric field is strong enough to pull some of the valence electrons into the conduction band by breaking their covalent bonds. These electrons then become free electrons which are available for conduction. A large number of such free electrons will constitute a large reverse current through the Zener diode and breakdown is said to have occurred due to the Zener effect.

2.3.1 Zener Characteristics

Working of Zener Diode

The zener diode is a silicon PN junction device, which diffuse from a rectifier diode, that it is operated in the reverse breakdown region. The breakdown voltage of zener diode is set by carefully controlling the doping level during manufacture.

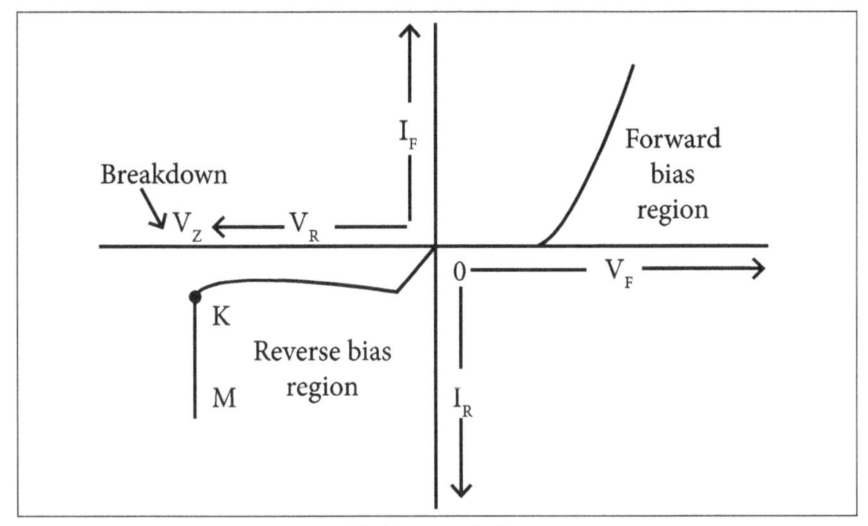

VI characteristics.

When a reverse voltage across a diode is increased, a critical voltage called breakdown voltage is reached at which the reverse current increases sharply.

The reverse breakdown of a PN junction may occur either due to avalanche or zener effect. The zener breakdown occurs when the electric field across the junction, produced due to reverse voltage, is sufficiently high. This electric field exerts a force on the electron in the outer most shell. This force is high that the electrons are pulled away from their parent nuclei and became free carriers. This ionisation, which occurs due to the electrostatic force of attraction is termed as zener effect. It causes an increase in the number of free carriers and hence an increases in the reverse current.

Zener Diode Regulator

Input Regulation

For input regulation, the input voltage (V_s) is kept fixed and the load resistance Rt varies. The variation of load resistance changes the current (I_L) through it, thereby changing the voltage (V_L) across it. When the load resistance decreases, the load current increases. This causes the zener current to decrease. As a result of this, the input current and the voltage drop across series resistance remains constant. Thus, the load voltage (V_L) is also kept constant.

On the other hand, if the load resistance increases, the load current decreases. As a result of this, the zener current increases. This again keeps the value of input current and voltage drop across series resistance as constant. Hence, the voltage remains constant.

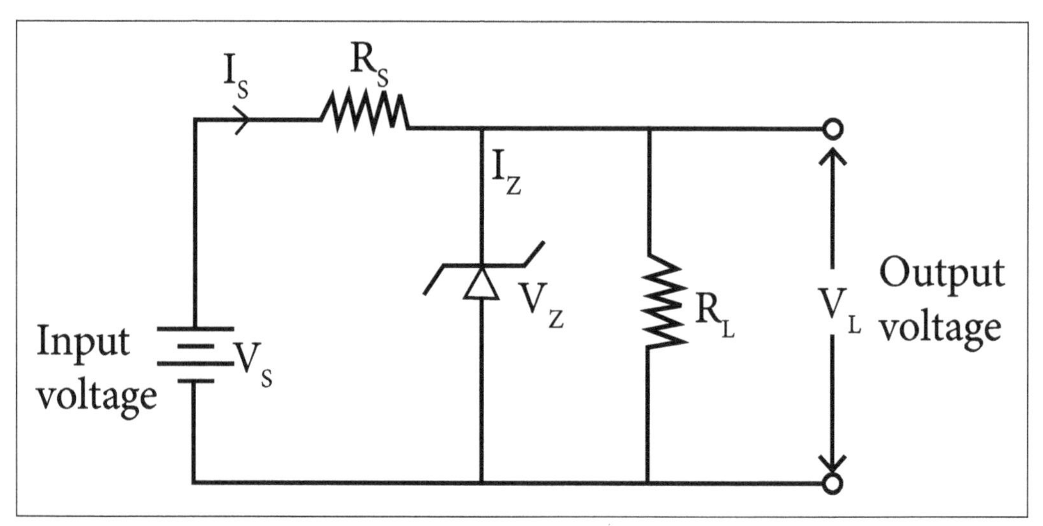

Voltage regulator using zener diode.

Here,

$V_s \rightarrow$ Input voltage

$I_s \rightarrow$ Input current

$R_s \rightarrow$ Series current limiting resistor

$I_z \rightarrow$ Zener current

$V_z \rightarrow$ Zener voltage

$R_L \rightarrow$ Load resistance

$V_L \rightarrow$ Output voltage

Output Regulation

Here the load resistance R_L is kept fixed and the input voltage (V_s) varies within the limits. As the input voltage increases, the input current (I_s) also increases. This increases, the current through the zener diode without affecting the load current (I_L).

Increase in the input current will also increase the voltage drop across the series resistance (R_s), thereby keeping the load voltage (V_L) as constant. On the other hand, the input current also decreases, if the input voltage is decreased. As a result of this, the current through zener will also decrease. Consequently, the voltage drop across series resistance will be reduced.

2.3.2 Tunnel Diode

A tunnel diode or Esaki diode is a type of semiconductor that is capable of very fast operation, well into the microwave frequency region, made possible by the use of the quantum mechanical effect called tunneling.

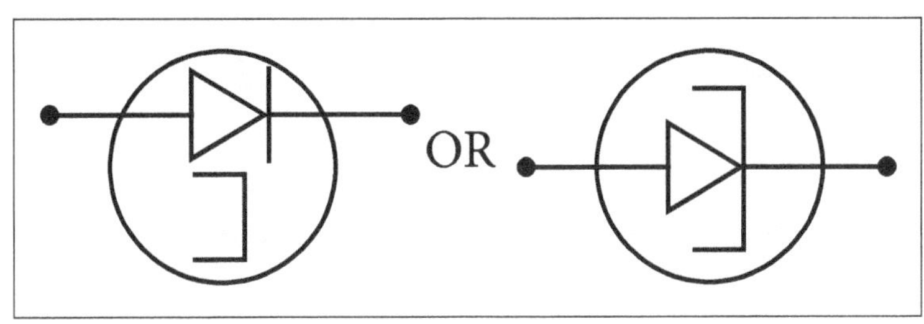

Tunnel Diode Symbol.

Tunnel Diode Construction and Working

If the concentration of impurity atoms is greatly increased in a normal PN junction, its characteristics are completely changed. This gives rise to a new type of device known as tunnel diode.

In a normally doped PN junction, the impurity concentration is of vary small values. With this amount of doping, the width of depletion layer is of the order of one micron. This makes a potential barrier at the junction, and that controls the flow of the charge carriers across the junction. The charge carriers cannot cross-over the potential barrier, unless it acquires the sufficient energy to overcome it.

However, when impurity concentration is increased, the width of the depletion layer is reduced to about 10 nanometer. Under certain conditions, the charge carriers will penetrate through the junction at the speed of light, even though they do not have enough energy to overcome potential barrier. As a result, a large forward current is produced, though the applied forward voltage is much less than 0.3 V.

The phenomenon of penetrating the charge carriers, directly through the potential barrier, instead of climbing over it is called tunneling. Hence, highly doped PN junction devices are called tunnel diodes. There diodes are usually made of germanium (Ge) or gallium arsenide.

Tunnel Diode Application

- It is used in relaxation oscillator.
- It is used as logic memory storage device.
- It is used as an ultra-high-speed switching devices.
- It is used as a microwave oscillator at frequencies in the order of 10 GHz.

2.3.3 Characteristics with the Help of Energy Band Diagrams

V-I Characteristics of a Tunnel Diode

Figure shows V-I characteristics of a tunnel diode. From the characteristics curve, we see that as the applied forward voltage is increased from zero, the current increases very rapidly till it reaches its maximum value known as peak current (I_p) as indicated by this point A. The corresponding values of the forward voltage is indicated by peak voltage (V_p). The value of this voltage is typically 65mV for germanium and 160m V for gallium arsenide tunnel diode.

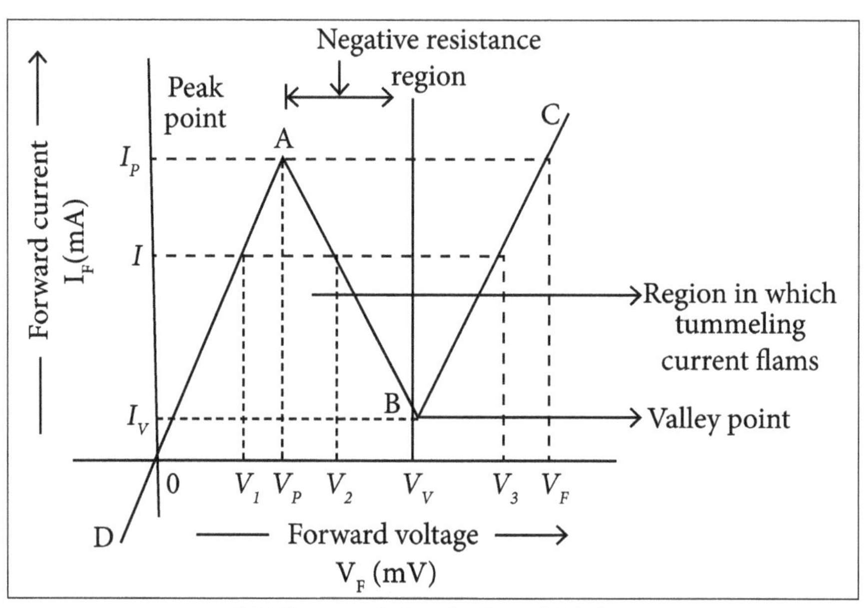

V-I Characteristics of a Tunnel Diode.

- If the forward voltage is further increased (i.e., beyond V_p), the current decreases, till it reaches its minimum value known as valley current (I_v) as indicated by the point B.
- As the voltage is further increased, the current increases in a usual manner as in a normal PN junction diode. It has been observed that the current gain reaches its peak value (i.e.) I_p as indicated by V_p as shown in the figure. For larger values of voltages, the current increases beyond these values.
- If the tunnel diode is reverse biased it acts like an excellent conductor, i.e., the reverse current increases with the increase in reverse voltage. It is indicated by the curve OD in the figure.
- Between the peak point A and valley point B, the current decreases with the increase in voltage, therefore, the tunnel diode possess a negative resistance in this region as indicated in the figure. This feature makes the tunnel diode useful in high frequency oscillators.

- For currents, whose values are between I_v and I_p, the curves is triple valued. It means that each current can obtained at three different applied voltages. It is indicated in the figure by the voltages V1, V2 and V3 for the current I. This multi valued feature makes the tunnel diode useful in pulse and digital circuit.

- The portion BC of the characteristics is similar to that of a forward-characteristics of a normal PN junction.

- The shaded region in the figure indicates the region in which the tunneling current flows through the device.

The characteristics curve has the following three current regions:

- Normal diode current: This is the 'normal' current that would flow through a PN junction diode.

- Tunneling current: This is the current that arises as a result of the tunneling effect.

Excess current: This is the third element of the current which contributes to the overall current within the diode. It results from the excess current that results from tunneling though bulk states in the energy gap and also means that the valley current does not fall to zero.

These three main components sum together to provide the overall level of current passed by the tunnel diode.

2.4 Varactor Diode, LED, Pin Diode and Photo Diode

Varactor Diode

Varactor diode is a semiconductor, voltage-dependent variable capacitor alternatively known as Varicap or Voltacap or Voltage-Variable Capacitor (VVC) or Tuning Diode.

Basically, it is just a reverse-biased junction diode whose mode of operation depends on its transition capacitance (C_T). A varactor diode is specially constructed to have high resistance under reverse bias. Capacitance for varactor diode are Pico farad range.

In a conventional diode, the depletion region exists between p-region andn-region as shown in figure.

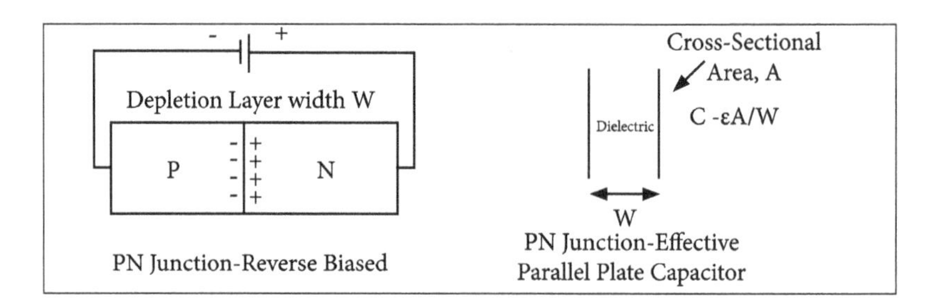

The holes are majority carriers in p-type semiconductor material and electrons are majority charge carriers in n-type semiconductor material. There is a neutral space charge region (junction) or the depletion between these two which do not have any majority carriers. There are two plates in the capacitor between which there is a dielectric material.

The p-region and n-region acts like the plates of capacitor while the depletion region acts like dielectric. Hence, there exists a capacitance at the pn junction called transition capacitance CT. It is denoted by the expression,

$$C_T = \frac{\varepsilon A}{W}$$

where,

C_T = Total Capacitance of the junction

C = Permittivity of the semiconductor material

A = Cross sectional area of the junction

W = Width of the depletion layer

Figure (a) shows an example of a cross section of a varactor with the depletion layer formed of a PN junction. This depletion layer can be made of a MOS or a Schottky diode. This is very important in CMOS and MMIC technology. Figure (b) shows the symbol of varactor diode.

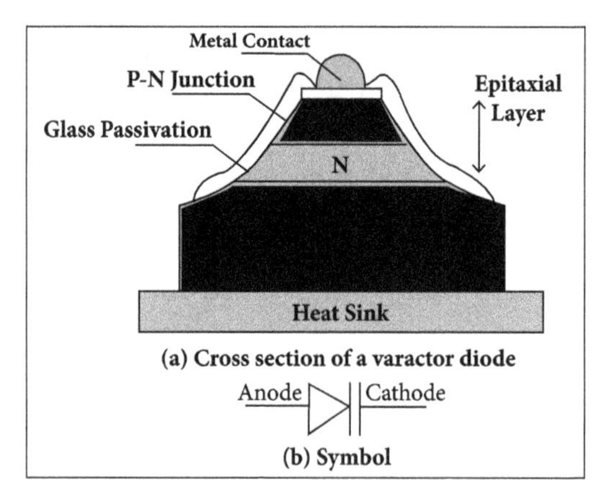

(a) Cross section of a varactor diode

(b) Symbol

Operation of Varactor Diode

Varactors are operated in reverse-biased state. No current flows but since the thickness of the depletion region changes with the applied bias voltage, the capacitance of the diode may be made to vary.Basically, the depletion region thickness is directly proportional to square root of the applied voltage. Its capacitance is inversely proportional to the depletion region thickness. Hence, the capacitance is inversely proportional to the square root of the applied voltage.

All diodes exhibit this phenomenon to some degree, but varactor diodes are manufactured specifically to exploit this effect and increase the capacitance. The capacitance may be controlled by the applied voltage. Curve between Reverse bias voltage VR across varactor diode and total junction capacitance CT is shown in the figure:

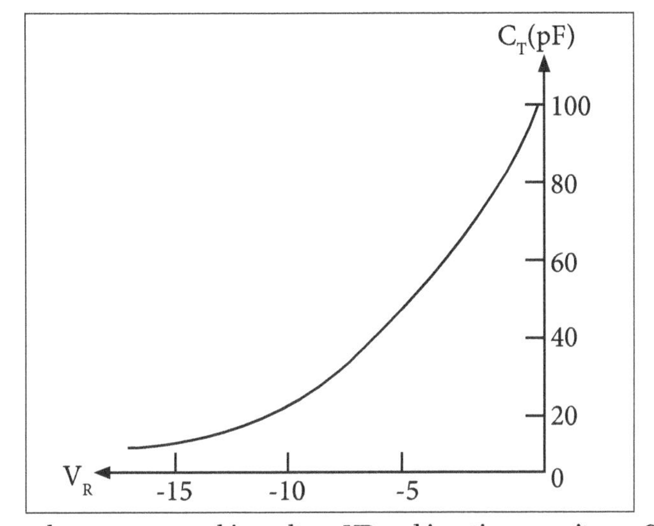

Curve between reverse bias voltage VR and junction capacitance CT.

Varactors may be of two types. The doping profile of the abrupt-junction diode and that of the hyper abrupt-junction diode. The abrupt-junction diode has uniform doping and a capacitive tuning ratio (TR) of 4:1. The hyper abrupt-junction diode has highest impurity concentration near the junction. It results in narrow depletion layer and larger capacitance. Also, changes in VR produce larger capacitance changes. Such a diode has a tuning range of 10:1 enough to tune a broadcast receiver through its frequency range of nearly 3:1.

Applications

- Tuned circuits.
- FM modulators.
- Parametric amplifiers.
- Television receivers.

- Adjustable band pass filters.
- Automatic frequency control devices.

Light Emitting Diode (LED)

The LED in an optical diode, which emits light when forward biased.

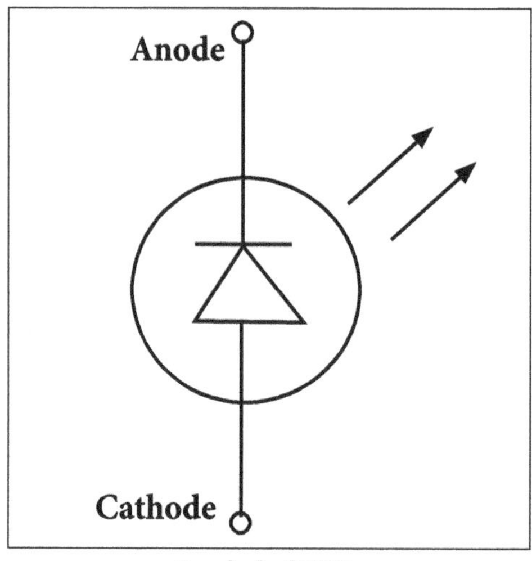

Symbol of LED.

The figure shows the symbol of LED which is similar to P-N junction diode apart from the two arrows indicating that the device emits light energy.

Basic Operation

Whenever a P-N junction is forward biased, the electrons cross the P-N junction from N-type semiconductor material and recombine with the holes in the P-type semiconductor material. The free electrons are in conduction band while the holes are present in the valence band. Hence, the free electrons are at higher energy level with respect to the holes.

When a free electron recombined with hole, it falls from conduction band to a valence band. Thus, the energy level associated with it changes from higher value to lower value.

The energy corresponding the difference between the higher level and lower level is released by an electron when it is traveling from the conduction band to the valence band. In normal diodes, this energy released is in the form of heat.

But LED is made up some special material which release this energy in the form of photons which emit light energy. Hence such diodes are called light emitting diodes. This process is called electroluminescence. The figure shows the basic principle of this process.

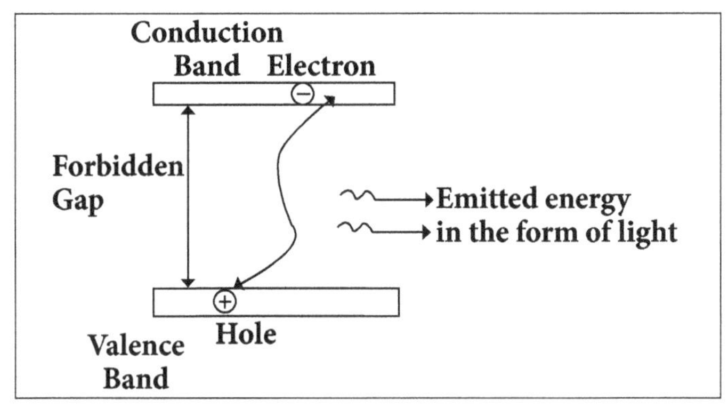

Process of Electroluminescence.

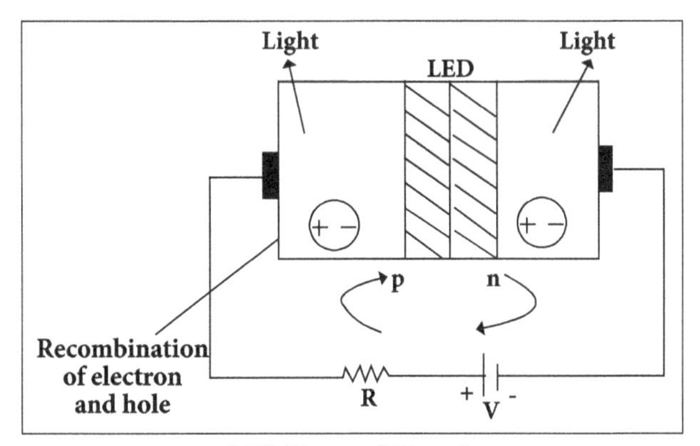

LED Forward Biased.

Operation of LED

The energy released in the form of light depends on the energy corresponding to the forbidden gap. This determines the wavelength of the emitted light. The wavelength determines the colour of the light and also determines whether the light is visible or invisible (infrared).

Various impurities are added during the doping process to control the wavelength and colour of the emitted light. For a normal silicon diode, the forbidden energy gap is 1.1 eV and wavelength of the emitted light is not visible. The infrared light is not visible.

Construction of LED

One of the methods used for the LED construction is to deposit the three semiconductor layers on the substrate as shown in the figure.

In between P-type and the N-type, there exists an active region. This active region emits light, when an electron and hole recombine. When the diode is forward biased, holes from the P-type and electrons N-type both get driven into the active region and when recombine, the light is emitted.

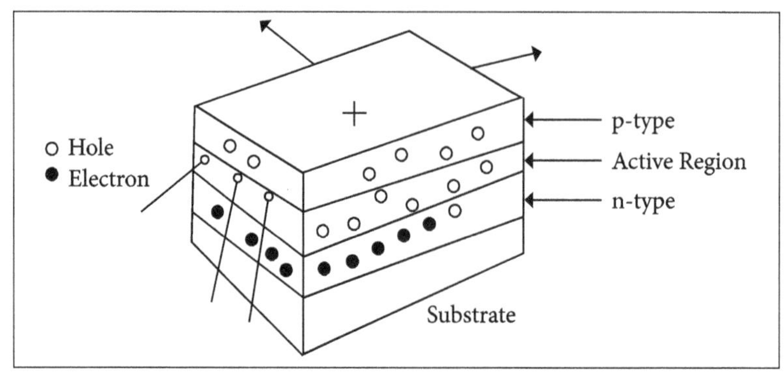

Construction of LED.

The symbol of LED indicating identification of anode and the cathode is shown in the figure:

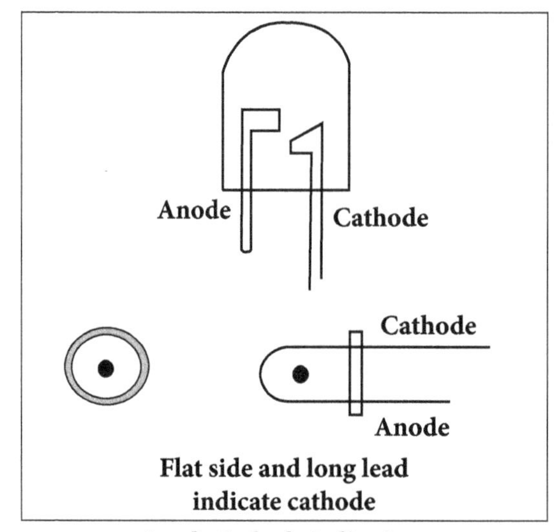

Anode-Cathode Indication.

Output Characteristics

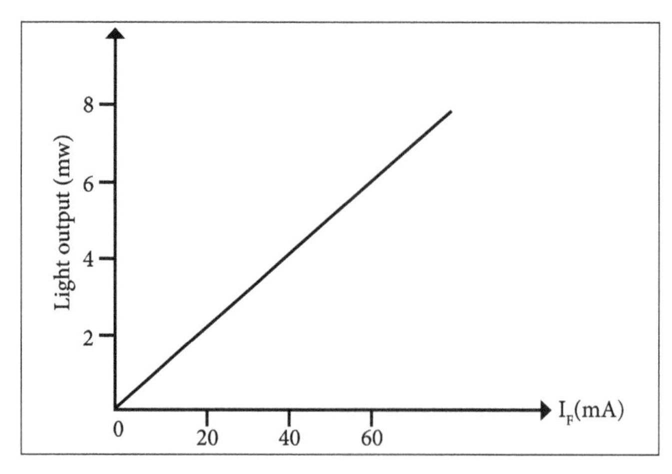

- The amount of power output that is translated into light is directly proportional to forward current (I_f).
- Greater the forward current I_f the greater is the output light.
- The graph of forward current and light output in mW.

Applications

- Used in display clocks, audio and video equipment, traffic lights.
- Used as light source in optical fiber communication.
- The low current LED's are useful in low power DC circuits, telecommunication indication, and portable equipment and keyboard indications.

Pin Diode

The pin diode has two narrow, but highly doped, semiconductor regions separated by a thicker, lightly-doped material called as the intrinsic region. As the name implies, one of the heavily doped regions is p-type material and the other is n-type. The same semiconductor material, generally silicon, is used for all the three areas. Silicon is used most often for its power-handling capability and also because it provides a highly resistive intrinsic region. The pin diode acts as an ordinary diode at frequencies up to about 100 megahertz, but above this frequency the operational characteristics change.

The large intrinsic region increases transit time of the electrons crossing the region. Above 100 megahertz, electrons start to accumulate in the intrinsic region. The carrier storage in intrinsic region makes the diode to stop acting as a rectifier and starts acting as a variable resistance.

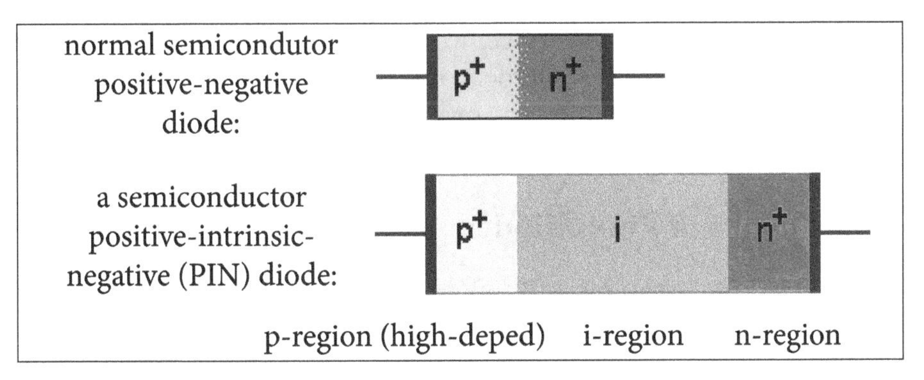

When the bias on a pin diode is modified, the microwave resistance changes from a typical value of 6 kilo ohms under the negative bias condition, and to about 5 ohms when the bias is positive. Hence, when the diode is mounted across a transmission line or waveguide, the loading effect is insignificant, but when the diode is reverse biased, and the diode presents no interference to the power flow. When the diode is forward biased, resistance drops to the approximate of 5 ohms and most of the power is reflected. In other words, the diode acts as a switch when mounted in parallel with the transmission line or waveguide. Various diodes in parallel may switch power in excess of the 150 kilowatts peak. The upper power limit is obtained by the ability of the diode to the power dissipated. The upper frequency limit is determined by the shunt capacitance of PN junction. Pin diodes with the upper limit frequencies in excess of 30 Gigahertz are available.

Photo Diode

A photodiode is a kind of light detector, which involves the conversion of light into voltage or current, based on the mode of operation of the device.

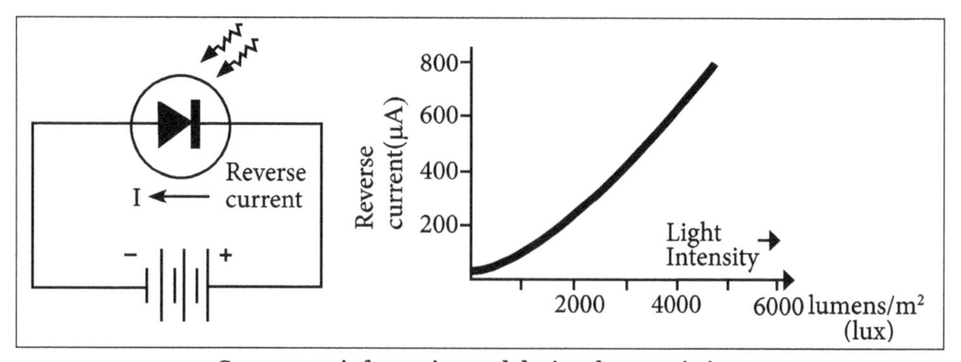

Component information and device characteristics.

It has built-in lenses and optical filters, and has small or large surface areas. Photodiodes have a slower response time, with an increase in their surface areas. Conventional solar cells are a typical photodiode with a large surface area, and are used for generating electric solar power, are a.

A photodiode is a semi-conductor device, with p-n junction and an intrinsic layer between p and n layers. It produces photocurrent by generating the electron-hole pairs, due to the absorption of light in intrinsic or in the depletion region. The photocurrent hence generated is proportional to the absorbed light intensity.

Working Principle of Photodiodes

When photons of energy greater than 1.1 eV hit the diode, electron-hole pairs are created. The intensity of photon absorption depends on the energy of photons i.e., the lower the energy of photons, the deeper the absorption is. This process is known as the inner photoelectric effect.

If the absorption occurs in depletion region of the p-n junction, these hole pairs are swept off from the junction - due to the built-in electric field of the depletion region. As a result, the holes move towards the anode and the electrons move towards the cathode, producing photocurrent. The sum of photocurrents and dark currents, which flow with or without light, is the total current passing through the photodiode. The sensitivity of the device can be increased by minimizing the dark current.

Modes of Operation

Photodiodes can be operated in various modes, which are as follows:

- Avalanche diode mode - Avalanche photodiodes are operated in high reverse bias condition, which allow multiplication of an avalanche breakdown to each photo-generated electron-hole pair. This results in internal gain within the photodiode, which gradually increases the responsively of the device.
- Photoconductive mode - The type of diode used in this mode is more commonly reverse biased. The application of reverse voltage increases the width of the depletion layer, which in turn reduces the response time and capacitance of the junction. This mode is very fast, and exhibits electronic noise.
- Photovoltaic mode – It is also known as the zero bias modes, in which a voltage is generated by the illuminated photodiode. It provides a very small dynamic range and non-linear dependence of the voltage produced.

Applications

Photodiodes find application in the following:

- Automotive devices.
- Medical devices.
- Cameras.
- Optical communication devices.
- Safety equipment.
- Position sensors.
- Surveying instruments.
- Bar code scanners.

3

Rectifiers and Regulators

3.1 Rectifiers and Regulators: Half Wave Rectifier and Ripple Factor

Half Wave Rectifier

The primary winding of the transformer is connected to an ac supply which induces an ac voltage across the secondary winding of the transformer.

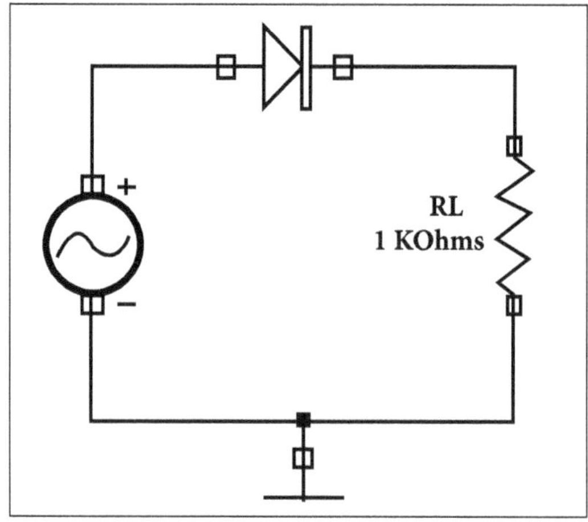

Half wave rectifier.

During the positive half cycle of the input, the polarity of the voltage across the secondary winding forward biases the diode. As a result, a current (I_L) flows through the load resistor (R_L). The forward biased diode offers a very low resistance and the voltage drop across it is very small. Thus the voltage appearing across the load is same as the input voltage at every instant.

During the negative half cycle of the input voltage, the polarity of the secondary voltage is reversed. Thus, the diode is reverse biased.

There is no current flow through the circuit and no voltage is developed across the resistor where all input voltage appears across the diode itself. Hence, when the input voltage is going through its positive half cycle, output voltage is same as the input voltage and during its negative half cycle, no voltage is available across the load.

Here, a unidirectional pulsating dc waveform is obtained as output. The process of removing one half of the input signal to establish a dc level is called as half wave rectification.

Peak Inverse Voltage

When the input voltage reaches its maximum value V_m during the negative half cycle, the voltage across the diode is maximum which is known as the peak inverse voltage.

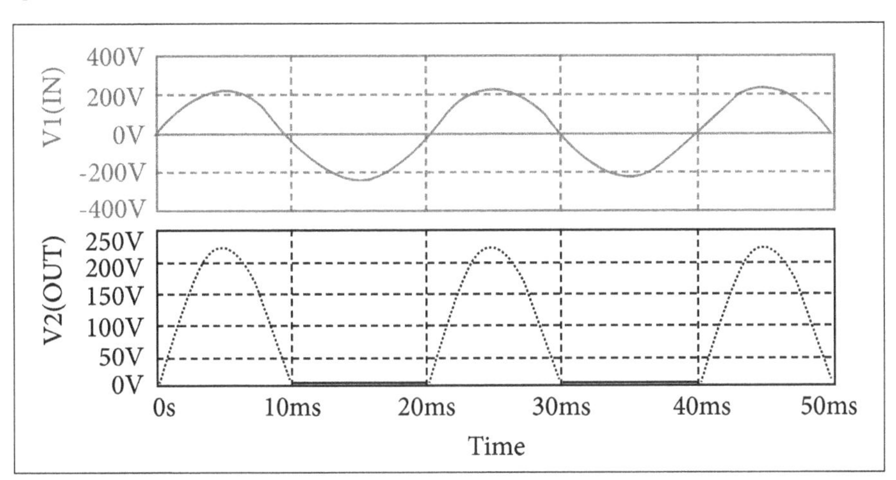

Output waveform of half wave rectifier.

Average Load Current (I_{dc})

$$I_{dc} = \frac{1}{2\pi}\int_{o}^{\pi} I_m \sin\omega t = \frac{I_m}{\pi}$$

$$i_L = I_m \sin\omega t$$

Where,

i_L – Load current

Efficiency (η)

DC power delivered to the load/AC input power from the transformer $= P_{dc} / P_{ac}$

DC power delivered to the load, $P_{dc} = I_{dc}^2 R_L = (I_{max} / p_i)^2 R_L$

AC power input to the transformer is given as,

P_{ac} = Power dissipated in diode junction + Power dissipated in load resistance (R_L)

$$= I_{rms}^2 R_F + I_{rms}^2 R_L = \{I_{max}^2 / 4\}[R_F + R_L]$$

Rectification efficiency is given as,

$$\eta = P_{dc} / P_{ac} = \{4/2\}\left[R_L / (R_F + R_L)\right] = 0.406/\{1 + R_F / R_L\}$$

The maximum efficiency of the half wave rectifier is 40.6% where R_F is neglected.

Form Factor

Form factor for any periodic wave is defined as the ratio of RMS and average value,

$$\text{i.e,}\ \text{F.F.} = \frac{\text{RMS value}}{\text{DC value}}$$

Therefore, form factor for the half wave rectifier is,

$$\text{F.F.} = \frac{\left(I_{m/2}\right)}{\left(I_{m/\pi}\right)} = \frac{\pi}{2}$$

Or

$$\text{F.F.} = 1.57$$

Average (d.c.) Load Voltage

We can determine the d.c. load voltage using the Ohm's law, i.e.,

$$\begin{aligned} V_{L_{dc}} &= I_{dc} \times R_L \\ &= \frac{V_m}{\pi\left(R_f + R_2 + R_L\right)} \times R_L \end{aligned}$$

If $R_L >>> \left(R_f + R_2\right)$, the general case, we get,

$$V_{L_{dc}} \simeq \frac{V_m}{\pi}$$

RMS Load Voltage

Similarly, the RMS value of load voltage can be find as,

$$\begin{aligned} V_{L_{rms}} &= I_{rms} \times R_L \\ &= \frac{V_m}{2\left(R_f + R_2 + R_L\right)} \times R_L \end{aligned}$$

Making the assumption, $R_L >>> (R_f + R_2)$, we get,

$$\therefore V_{L_{rms}} \simeq \frac{V_m}{2}$$

3.1.1 Ripple Factor

The rectifier output consists of a.c. as well as d.c. The ripple factor measures the percentage of a.c. component in the rectified output. The ideal value of ripple factor should be zero, i.e., output should be pure d.c.

Ripple factor is defined as,

$$\text{Ripple factor, } \gamma = \frac{\text{RMS value of the a.c. component of output}}{\text{d.c. or Average value of output}}$$

As the instantaneous value of a.c. fluctuations is measured with respect to the d.c. level (i.e., the difference of instantaneous total value and the d.c. value). Thus, the instantaneous a.c. value is given as,

$$i_{ac} = i - I_{dc}$$

Therefore, the RMS value oa.c. components is given as,

$$I_{ac_{rms}} = \left[\frac{1}{2\pi}\int_0^{2\pi}(i - I_{dc})^2 \, d(\omega t)\right]^{1/2}$$

$$= \left[\frac{1}{2\pi}\int_0^{2\pi}(i^2 + I_{dc}^2 - 2\, i\, I_{dc}) d(\omega t)\right]^{1/2}$$

$$= \left[\frac{1}{2\pi}\int_0^{2\pi} i^2\, d(\omega t) + \frac{1}{2\pi}\int_0^{2\pi} I_{dc}^2\, d(\omega t) - 2\, I_{dc}\int_0^{2\pi} i\, d(\omega t)\right]^{\frac{1}{2}}$$

$$= \left[I_{rms}^2 + I_{dc}^2 - 2 I_{dc}^2\right]^{\frac{1}{2}}$$

$$I_{ac_{rms}} = \left[I_{rms}^2 - I_{dc}^2\right]^{\frac{1}{2}}$$

Therefore, ripple factor,

$$\gamma = \frac{I_{ac_{rms}}}{I_{dc}} = \frac{\left[I_{rms}^2 - I_{dc}^2\right]^{\frac{1}{2}}}{I_{dc}}$$

$$= \left[\left(\frac{I_{rms}}{I_{dc}}\right)^2 - 1\right]^{\frac{1}{2}}$$

$$\gamma = \frac{I_{ac_{rms}}}{I_{dc}} = \frac{\left[I_{rms}^2 - I_{dc}^2\right]^{\frac{1}{2}}}{I_{dc}}$$

$$= \left[\left(\frac{I_{rms}}{I_{dc}}\right)^2 - 1\right]^{\frac{1}{2}}$$

$$\gamma = \left[(\text{F.F.})^2 - 1\right]^{\frac{1}{2}}$$

For half wave rectifier, ripple factor can be calculated by substituting the value of form factor into the above equation, therefore,

$$\gamma = \left[(1.57)^2 - 1\right]^{\frac{1}{2}}$$

Or

$$\gamma = 1.21$$

Advantages

- Transformer is not compulsory.
- Circuit is suitable for high voltage applications because PN is less, i.e., equal to V_m.
- Core saturation does not take place.

Disadvantage

- The only disadvantage of this circuit is the need of four diodes, but it is not major one compared to the advantages of the circuit.

3.2 Full Wave Rectifier and Harmonic Components in a Rectifier Circuit

Full Wave Rectifier (with and without Transformer)

The full wave rectifier is a circuit which allows a unidirectional current flow through the load during the entire input cycle. There are two types of full wave rectifier namely center-tapped and bridge rectifier.

Bridge Rectifier

The need for center-tapped power transformer is eliminated in the bridge rectifier.

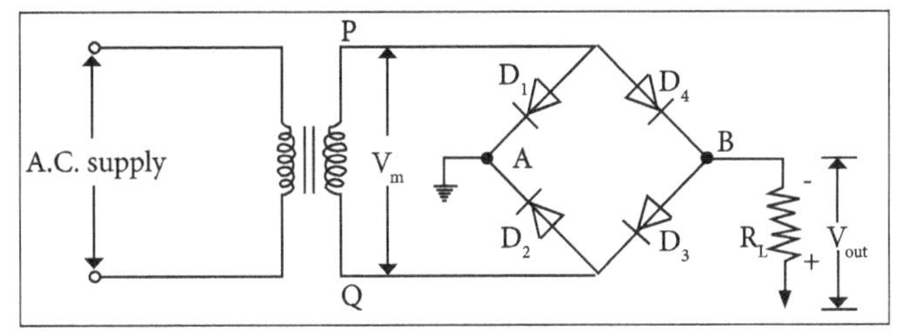

Circuit diagram of bridge rectifier.

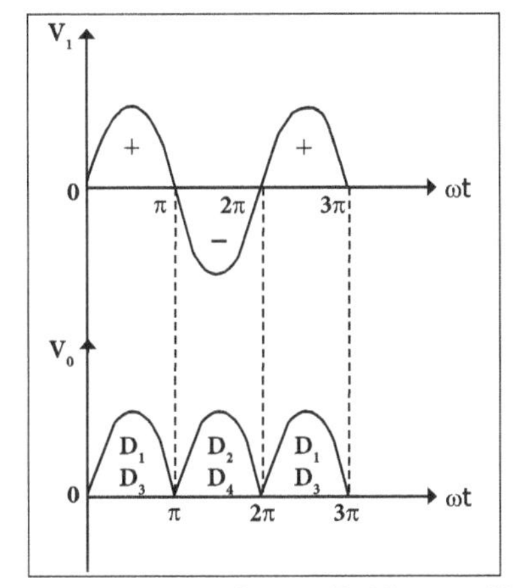

Input-output waveform of bridge rectifier.

It contains four diodes namely D_1, D_2, D_3 and D_4 which are connected to form a bridge. The a.c supply to be rectified is applied to the diagonally opposite ends of the bridge through the transformer between the load (R_L).

During the positive half-cycle (P-positive and Q-negative), D_1 and D_3 are in forward bias, D_2 and D_4 are in reverse bias.

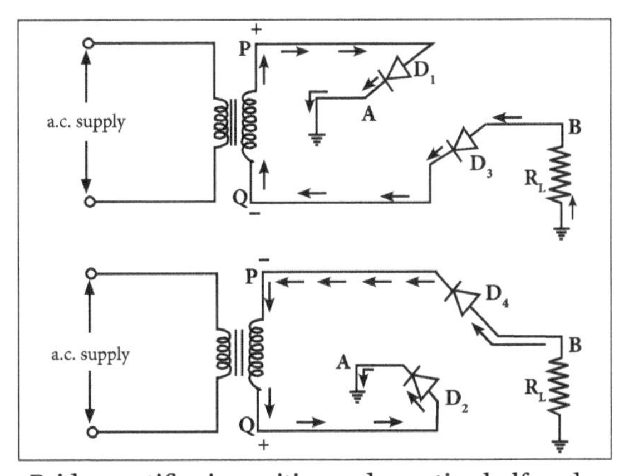

Bridge rectifier in positive and negative half-cycle.

During the negative half-cycle, D_1 and D_3 are in reverse bias and D_2 and D_4 are in forward bias conditions. These two diodes will be in series with the load (R_L) where the current flows from A to B through the load in the same directions and d.c output is obtained across the load (R_L).

Transformer Utilization Factor (TUF)

$$TUF = \frac{\text{d.c power delivered to the load}}{\text{a.c rating of the transformer}}$$

$$= \frac{P_{dc}}{P_{ac} \text{ rated}}$$

$$= \frac{\left(\frac{2I_m}{\pi}\right)^2 .R_I.\left(\frac{2I_m}{\pi}\right)^2 .R_I.}{\frac{V_m}{\sqrt{2}}.\frac{I_m}{\sqrt{2}} \quad \frac{I_m^2}{2}.R_I} \qquad [\because V_m = I_m R_I]$$

$$TUF = \frac{8}{\pi^2}$$

$$TUF = 0.81 \text{ or } 81\%$$

Peak Inverse Voltage (PIV)

It is defined as the maximum reverse voltage that a diode can withstand without destroying the junction.

$$PIV = 2V_m$$

The same procedure is repeated when D_1 is OFF and D_2 is ON.

Center-tapped Full Wave Rectifier

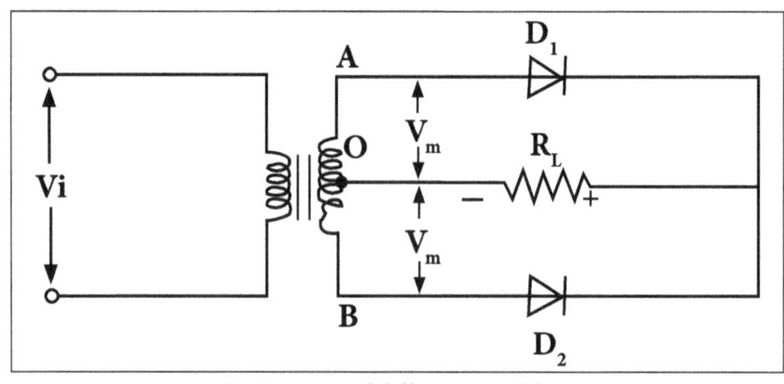

Center-tapped full wave rectifier.

In full wave rectification, current flow through the load is in the same direction for both half-cycles of input a.c voltage. It uses center-tapped transformer which provides equal voltage above and below the center-tapped for both the half-cycles.

The center-tap on the secondary winding of a transformer is taken as zero voltage or ground reference point. The circuit uses two diodes which are connected to the center-tapped secondary winding of the transformer. Diode D_1 utilizes the ac voltage that appears across the upper half (OA) of the secondary winding for rectification while diode D_2 uses the lower half winding (OB).

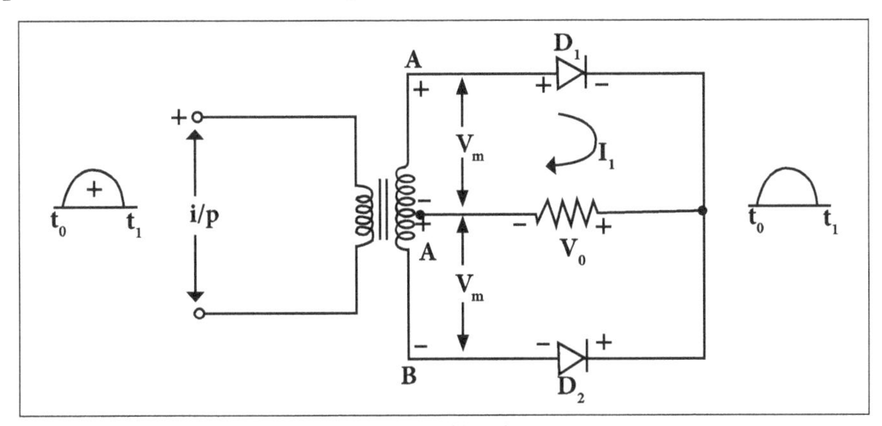

(+)ve half cycle.

During the positive half cycle of secondary voltage, the end A of the secondary winding becomes positive and end B becomes negative as shown in the above figure. This makes the diode D_1 forward biased and D_2 reverse biased. Therefore, diode D_1 conducts while diode D_2 does not conduct. The conventional current flow takes place through diode D_1, load resistor R_L and upper half of secondary winding.

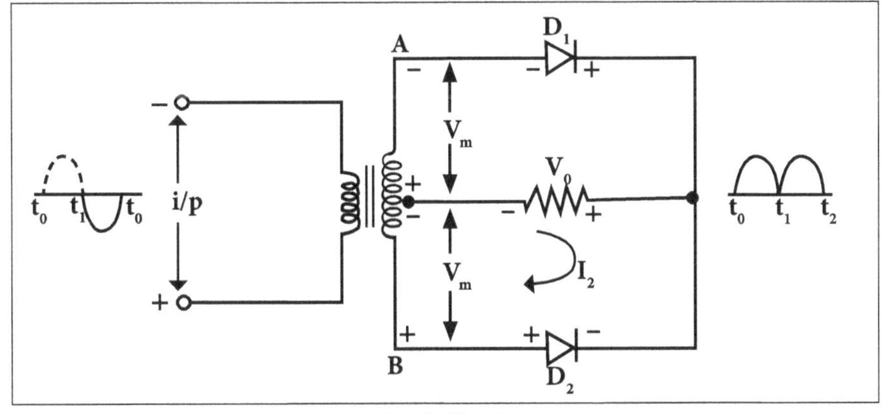

(-)ve half cycle.

During the negative half cycle, the end A of the secondary winding becomes negative and the end B becomes positive as shown in the above figure. Therefore, diode D_2 conducts while diode D_1 does not conduct. The conventional current flow takes place through diode D_2, load resistor R_L and lower half winding. The current in the load resistor R_L is in the same direction for both half-cycle of input a.c. voltage.

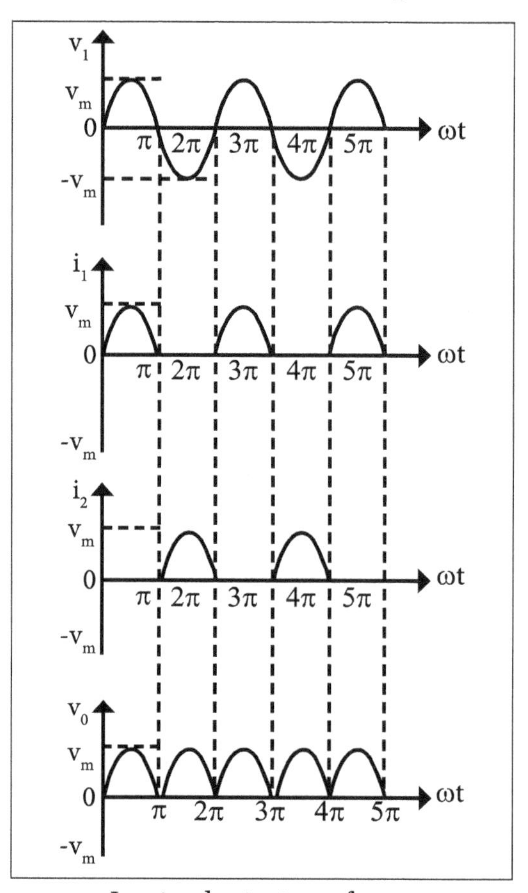

Input and output waveforms.

Average or dc Value

$$I_{dc} = \frac{\text{Area under the curve over a half cycle}}{\text{Base or period}}$$

$$= \frac{1}{\pi}\int_0^{\pi} I_L . d\omega t$$

$$I_{dc} = \frac{1}{\pi}\int_0^{\pi} I_m \sin\omega t \cdot d\omega t$$

$$= \frac{I_m}{\pi}[-\cos\,\omega t]_0^{\pi}$$

$$= \frac{I_m}{\pi}[\cos\pi - \cos\theta]$$

$$I_{dc} = \frac{2I_m}{\pi} \text{ Similarly } V_{dc} = \frac{2V_m}{\pi}$$

Form Factor

As we have from the definition of Form Factor that,

$$F.F. = \frac{\text{RMS value}}{\text{Average value}}$$

$$F.F. = \frac{\left(I_m / \sqrt{2}\right)}{\left(2\,I_m / \pi\right)} = \frac{\pi}{2\sqrt{2}}$$

Or $F.F. = 1.11$

RMS Value of Load Current

$$I_{rms} = \sqrt{\frac{1}{\pi}\int_0^{\pi}(I_m \sin\omega t)^2 . d\omega t}$$

$$= \sqrt{\frac{I_m^2}{\pi}\int_0^{\pi}\left(\frac{1-\cos 2\omega t}{2}\right)d\omega t}$$

$$= \sqrt{\frac{I_m^2}{2\not{\pi}} \cdot \not{\pi}}$$

$$I_{rms} = \frac{I_m}{\sqrt{2}} \qquad V_{rms} = \frac{V_m}{\sqrt{2}}$$

RMS Load Voltage

The RMS value of load voltage is given by,

$$V_{L_{rms}} = I_{rms} + R_L$$

Substituting the value of I_{rms}, we get as,

$$V_{L_{rms}} = \frac{I_m}{\sqrt{2}} \times R_L$$

Or $V_{L_{rms}} = \frac{1}{\sqrt{2}} \frac{V_m}{(R_L + R_f + R_2)} \times R_L$

$$\frac{1}{\sqrt{2}} \frac{V_m}{1+\left(\frac{R_f + R_2}{R_L}\right)}$$

Assuming $R_f + R_2 << R_L$, then

$$V_{L_{rms}} \simeq \frac{V_m}{\sqrt{2}}$$

Ripple Factor

$$\gamma = \frac{V_{r.rms}}{V_{dc}} = \frac{\sqrt{V_{rms}^2 - V_{dc}^2}}{V_{dc}}$$

$$= \sqrt{\left(\frac{V_m}{\sqrt{2}} \cdot \frac{\pi}{2V_m}\right)^2 - 1}$$

$$\gamma = 0.483 \text{ or } 48.3\%$$

Therefore, the quality of the d.c. voltage obtained from full-wave rectifier is better than that of half wave rectifier, because the ripples present in the rectified output of full wave rectifiers are just 48% which are much less than the 121% of half wave rectifier (HWR).

Efficiency

$$\eta = \frac{P_{dc}}{P_{dc}} = \frac{I_{dc}^2 \cdot R_L}{I_{rms}^2 (R_L + R_s + R_f)}$$

$$I_{dc} = \frac{2I_m}{\pi} \text{ and } I_{rms} = \frac{I_m}{\sqrt{2}}$$

$$\eta = \frac{\left(\frac{2\,I_m}{\pi}\right)^2 . R_L}{\left(\frac{I_m}{\sqrt{2}}\right)^2 (R_L + R_s + R_f)}$$

$$= \frac{8}{\pi^2} \cdot \frac{R_L}{(R_L + R_s + R_f)}$$

If $(R_s + R_f) << R_L$, then $\eta = 0.812$ or 81.2%

Thus the maximum theoretical rectification efficiency, which can be achieved with a full wave rectifier is 81.06%. It is much larger than that of half wave rectifier circuit, i.e., 40.53%.

Peak Inverse Voltage (PIV)

It is defined as the maximum voltage that a diode can withstand under reverse biased condition.

$$PIV = 2V_m$$

Transformer Utilization Factor (TUF)

$$TUFs = \frac{P_{dc}}{P_{dc} \cdot rated}$$

$$= \frac{I_{dc}^2 R_L}{V_{rms} \cdot I_{rms}} = \frac{(2I_m/\pi)^2 \cdot R_L}{\frac{V_m}{\sqrt{2}} \cdot \frac{I_m}{\sqrt{2}}}$$

$$= \frac{8}{\pi^2} = 0.811 \quad (or) \quad 81.1\%$$

Advantages

- Output voltage and transformer efficiency are higher.
- Transformer utilization factor is high.
- Ripple factor is low.

Disadvantages

- Additional diodes and bulky transformers are needed which increases the cost.
- Output voltage is half of the secondary voltage.
- PIV of diode is high (i.e., $2V_m$).

3.2.1 Harmonic Components in a Rectifier Circuit

An analytic representation of the output of the single-phase half-wave rectifier is obtained in terms of a Fourier series expansion.

This series representation is given as,

$$i = b_o + \sum_{k-1}^{\infty} b_k \cos k\alpha + \sum_{k-1}^{\infty} a_k \sin k\alpha \qquad ...(1)$$

Where,

$$\alpha = \omega t$$

The coefficients that appear in the series are given in the integrals as,

$$b_o = \frac{1}{2\pi} \int_0^{2r} i \, d\alpha$$

$$b_k = \frac{1}{\pi}\int_0^{2r} i \cos k\alpha \, d\alpha$$

$$a_k = \frac{1}{\pi}\int_0^{2r} i \sin k\alpha \, d\alpha \qquad ...(2)$$

The constant term b_0 that appears in this Fourier series is the average or dc value of the current.

The explicit expression for the current in a half-wave rectifier circuit is obtained by performing the indicated integrations.

$$i = I_m\left[\frac{1}{\pi} + \frac{1}{2}\sin\omega t - \frac{2}{\pi}\sum_{k=2,4,6...}\frac{\cos k\omega t}{(k+1)(k-1)}\right] \qquad ...(3)$$

Where,

$I_m = E_m / (r_p + R_t)$

E_m = Peak potential of the transformer

The lowest angular frequency that is present in the above expression is the primary source. For this single term of frequency ω, all other terms that appear in the expression are even-harmonic terms.

Fourier series representation of the output of the full-wave rectifier is derived from the above equation (3). Thus, the full-wave circuit has two half-wave circuits which are arranged in such a way that one circuit is operating during the interval when the other is not operating where the currents are given by the expression $i_2(\alpha) - i_1(\alpha + \pi)$. The total load current $(i = i_1 + i_2)$ is given as,

$$i = I_m\left[\frac{2}{\pi} - \frac{4}{\pi}\sum_{k=2,4,6...}\frac{\cos k\omega t}{(k+1)(k-1)}\right] \qquad ...(4)$$

Where,

$I_m = E_m / (R_t + r_p)$

E_m = Maximum value of the transformer potential measured at the center tap

A comparison of equations (3) and (4) indicates that the fundamental angular-frequency term is eliminated in the full-wave circuit and the lowest harmonic term in the output is 2ω which is a second-harmonic term.

Fourier series representation of the half-wave and full-wave circuits uses gas diodes which is more complex. Conduction begins at some small angle and ceases at the angle $\pi - \phi_o$ when the breakdown and the extinction potentials are equal.

Since these angles are small under normal operating conditions, it is assumed that equations (3) and (4) are applicable for circuits with vacuum or gas diodes. Fourier series representation of the output of a controlled rectifier is possible though the result is complex.

Such controlled rectifiers are used in services in which the ripple is not of major concern and as a result, no detailed analysis will be undertaken.

Problems

1. A half wave rectifier uses a diode with an equivalent forward resistance of 0.3 kΩ. If the input a.c. voltage is 10 V (rms) and the load is a resistance of 2.0 Ω, let us calculate I_{dc} and I_{rms} in the load.

Solution:

Given that:

$$V_{rms} = 10\,V$$

$$r_f = 0.3\,\Omega$$

$$R_L = 2.0\Omega$$

So, $V_m = \sqrt{2}\,V_{rms} = 10\sqrt{2}\,V$

Peak value of current in load, I_{max} or I_m

$$= \frac{V_m}{R_L + r_f} = \frac{10\sqrt{2}}{2 + 0.3} = 6.15\,A$$

dc output current, $I_{dc} = \frac{I_m}{\pi} = \frac{6.15}{\pi} = 1.958\,A.$

RMS value of output current,

$$I_{rms} = \frac{I_m}{2} = \frac{6.15}{2} = 3.075\,A.$$

2. Let us find the ripple 5 V on average of 50 V.

Solution:

Given:

$$V_{av} \text{ or } V_{dc} = 50\,V$$

$$V_{rms} = 5\,V$$

We know that ripple factor,

$$\gamma = \frac{V_{rms}}{V_{dc}} = \frac{5}{50} = 0.10$$

3. A single phase full-wave rectifier uses two diodes, the internal resistance of each being 20 Ω. The transformer rms secondary voltage from center tap to each end of secondary is 50 V and Load Resistance is 980 Ω.

Let us find:

- The mean load current.
- rms load current.
- Output efficiency.

Solution:

Given that:

$$V_{rms} = 50\,V$$
$$r_f = 20\Omega$$
$$R_L = 980\Omega$$

So, $V_m = \sqrt{2}\,V_{rms} = 50\sqrt{2}$

Mean load current

$$I_{dc} = \frac{2I_m}{\pi} = 2\frac{2.V_m}{\pi(R_L + r_f)} = \frac{2}{\pi}.\frac{50\sqrt{2}}{(980+20)}$$

$$= 45\,mA. \quad \text{Ans.}$$

rms load current

$$I_{rms} = \frac{I_m}{\sqrt{2}} = \frac{V_m}{\sqrt{2}(R_L + r_f)} = \frac{50\sqrt{2}}{\sqrt{2}(980+20)} = 50\,mA.$$

Output efficiency,

$$\eta = \frac{.812}{1+\frac{r_f}{R_l}} = \frac{.812}{1+\frac{20}{980}} = 79.58\%.$$

4. For a center-tapped transformer full wave rectifier, the turn ratio of the transformer between primary and half the secondary is 4:1. The resistance of each half of the secondary is 2Ω and the forward resistance of each diode is 1Ω.

Let us calculate:

- Average load current.
- Average load voltage at no load.
- Average load voltage at full load.
- Percentage load regulation.
- Rectification efficiency.

Solution:

Given:

$$R_f = 1\ \Omega,\ R_2 = 2\ \Omega,\ R_L = 1k\ \Omega$$
$$N_1 : N_2 = 4:1$$

$$I_{L_{DC}} = \frac{2V_m}{\pi(R_2 + R_f + R_L)}$$

Where,

$$V_m = \sqrt{2} \times V_{2_{rms}}$$

$$V_{2_{rms}} = \frac{N_2}{N_1} \times V_{1_{rms}} = \frac{1}{4} \times 240$$

Or,

$$V_{2_{rms}} = 60\,V$$

$$\therefore\ V_m = \sqrt{2} \times V_{2_{rms}}$$

$$= \sqrt{2} \times 60$$

$$V_m = 84.85\,\text{volt}$$

Therefore,

$$I_{L_{DC}} = \frac{2 \times 84.85}{\pi(2 + 1 + 1000)}$$

$$I_{L_{DC}} = 53.86 \text{ mA.}$$

Average load voltage at no load,

$$V_{NL} = \frac{2V_m}{\pi} = \frac{2 \times 84.85}{\pi}$$

$$V_{NL} = 54.02 \text{ V.}$$

Average load voltage at full load,

$$V_{L_{DC}} = V_{L_{DC}} \times R_L = 53.86 \times 1$$

$$V_{L_{DC}} = 53.86 \text{ volt.}$$

Load regulation

$$\% \text{ load regulation} = \frac{V_{NL} - V_{FL}}{V_{FL}} \times 100\%$$

$$= \frac{54.02 \quad 53.86}{53.86} \times 100$$

$$\therefore \qquad \% \text{ load regulation} = 0.297\%$$

Rectification Efficiency (η)

$$\text{Input a.c. power } P_{i_{AC}} = I_{rms}^2 (R_2 + R_f + R_I)$$

$$= \frac{I_m^2}{2}(R_2 + R_f + R_I)$$

$$\text{But, } I_m = \frac{V_m}{(R_L + R_2 + R_f)}$$

$$\Rightarrow P_{i_{AC}} = \frac{V_m^2}{2(R_L + R_2 + R_f)} = \frac{(84.85)^2}{2(1000 + 2 + 1)}$$

$$P_{i_{AC}} = 3.6 \text{ watts}$$

and output d.c. power,

$$P_{O_{DC}} = I_{L_{DC}}^2 R_L$$

$$=(53.86\times10^{-3})^2\times1000$$

$$=2.9 \text{ watts}$$

As we know, rectification efficiency,

$$\eta=\frac{P_{O_{DC}}}{P_{i_{AC}}}=\frac{2.9}{3.6}$$

$$\eta=80.56\%$$

5. For a full wave rectifier circuit (with center tapped transformer), let us find the average, RMS and peak values of currents through the diode, if the voltage across half of secondary is 15 sin 314 t. Let us also calculate the PIV of diode.

Solution:

Given for a outer-tapped full wave rectifier circuit,

$$v_i=15\sin 314\,t$$

$$\Rightarrow\therefore\quad V_m=15\,V$$

Average current through the diodes

$$I_{D_{\rightarrow}}=I_{dc}=\frac{I_m}{\pi}$$

But,

$$I_m=\frac{V_m}{R_L}\ \text{(as } R_f \text{ and } R_2 \text{ are assumed zero)}$$

$$\Rightarrow\ I_{D_{av}}=\frac{15}{\pi\times1}=4.77\,mA$$

$$I_{D_{av}}=4.77\,mA.$$

RMS current:

$$I_{D_{rms}}=\frac{I_m}{2}$$

as one diode conducts only for one half cycle,

$$\Rightarrow I_{D_{rms}}=\frac{15}{2\times1}=7.5\ mA$$

$$\therefore\ I_{D_{rms}} = 7.5\,\text{mA}$$

Peak diode current:

$$I_{D_{peak}} = I_m = \frac{V_m}{R_L}$$

or

$$I_{D_{peak}} = \frac{15}{1}$$

or

$$I_{D_{peak}} = 15\,\text{mA}.$$

Peak inverse voltage

As we know the PIV, appears across the diode in center-tapped full wave rectifier is given by,

$$\text{PIV} = 2\,V_m = 2 \times 15$$

or

$$\text{PIV} = 30\,\text{V}.$$

3.3 Filters and Comparison of Various Filter Circuits in Terms of Ripple Factors

Filter is an electronic circuit which is composed of a capacitor, inductor or combination of both and it is connected between the rectifier and the load to convert pulsating dc to pure dc. The different types of filters are given as,

- Inductor filter.
- Capacitor filter.
- LC or L-section filter.
- CLC or ∏-section filter.

3.3.1 Inductor Filter and Capacitor Filter

This type of filter is also called as choke filter. It consists of an inductor (L) which is inserted between the rectifier and the load resistance (RL). The rectifier contains a.c components as well as d.c components.

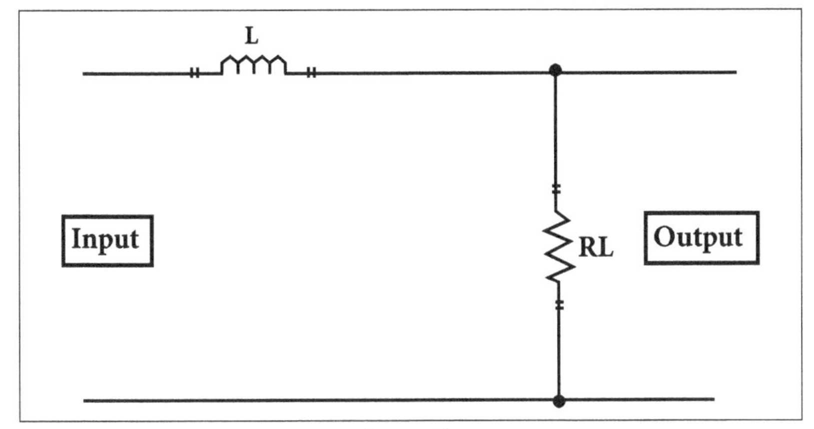

Inductor filter.

When the output passes through the inductor, it offers a high resistance to the a.c component and no resistance to d.c components. Therefore, a.c components of the rectified output is blocked and only d.c components reaches the load.

Capacitor Filter

Shunt Capacitor Filter

This is the most simple form among all the filter circuits and in this arrangement a high value capacitor C is placed directly across the load terminals as illustrated in the below figure. During the conduction period capacitor gets charged and stores up energy in the electrostatic field and discharges through the load resistance RL during the non-conduction period.

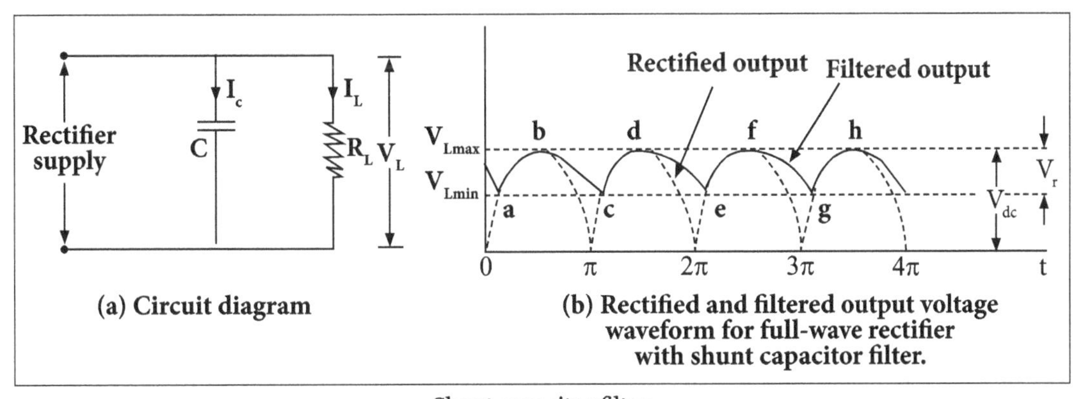

Shunt capacitor filter.

In this process, the time duration during which current flows through the load resistance gets prolonged (tedious) and a.c. component or ripples get considerably reduced. Here it is very interesting to note that the capacitor C get charged to the peak value of input voltage quickly because charging time constant and is almost zero. It is so because there is no resistance (except the negligible forward resistance of diode) in the charging path.

Series Inductor Filter

Circuit diagram of series inductor filter is shown in the below figure (a). In this arrangement (filter) a high value inductor (choke) L is connected in series with the load resistance and rectifier supply (output). Due to the unique property of inductor, its filtering action depends upon its property of opposing any change in the current flowing through it when the output current of the rectifier increases above a certain value, energy is stored in it in the form of magnetic field and energy is given up when the output current falls below the average value. So, placing a inductor in series with the rectifier output and load, any sudden change in current that might have occurred in the circuit without an inductor is smoothed out by the presence of the inductor L.

Again, the inductor uses the property that it opposes any change of current flowing through it. Inductor stores energy when the input signal is positive and delivers the energy when the input signal is negative. So a lower magnitude of voltage is obtained at the output. As the input decreases from zero, the inductor again opposes the decrease in current and supply energy to the load. Thus, the current at the load remains continuous with reduced pulsating nature.

We know that the choke offers high impedance to the a.c. component, but offers almost zero resistance to the derived d.c. components. Thus ripples are removed up to a large extent. Nature of the output voltage without filter and with choke filter are shown in the below figure (b).

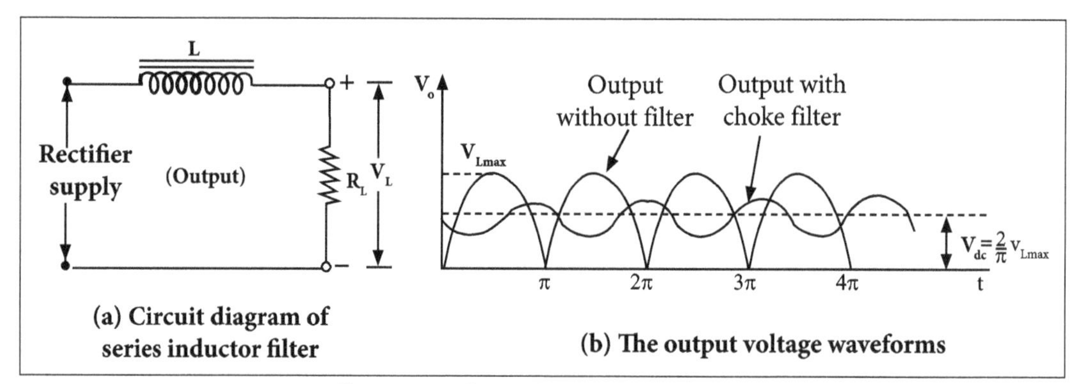

Full wave rectifier with series inductor filter.

3.3.2 L-section Filter and π- section Filter

The former reduces both the peak and effective values of the output current and output voltage while the later reduces the ripple voltage but at the same time increases the diode current which may cause the diode to be damaged. So, individual application of these filters in the power supply system adds certain disadvantages.

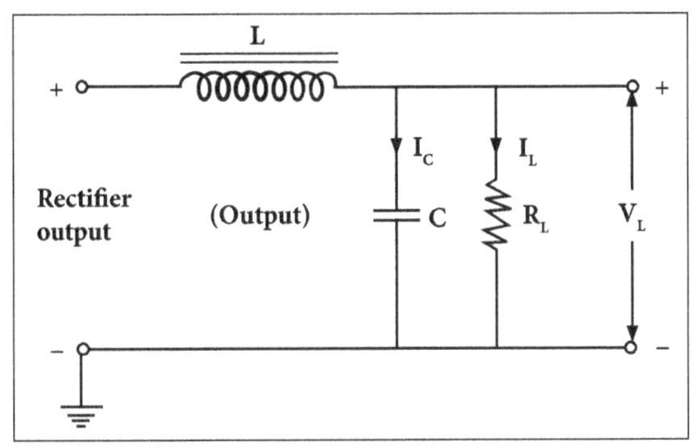

Circuit diagram of L-section filter.

In order to avoid the drawbacks of these two filter circuits and to add up their advantages, another type of filter is introduced which is referred to as the choke input or L-section filter. The following figure illustrates the arrangement of combination of inductor and capacitor in the filter circuit.

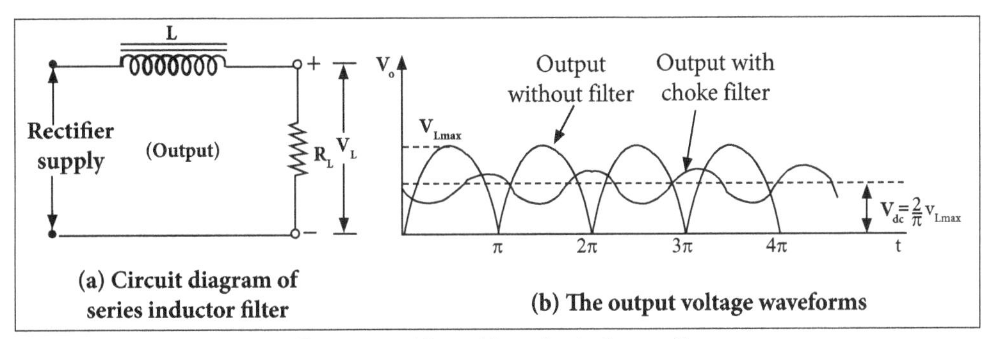

Full wave rectifier with series inductor filter.

The circuit diagram in the above figure consists of a inductor in series and capacitor in shunt with a load RL. The name L-section of this filter circuit has been derived from the basic 'L' shaped structure of the circuit.

The input is fed through the inductor so it is also known as the choke—input filter. Here, the inductor plays its role as a current smoothing element and capacitor as the voltage stabilizing element.

Here, it is notable that several L-sections in cascade can be connected in order to get more smooth filter output.

The choke L on the input side of the circuit easily allows d.c. to pass ($2\pi f L = 0$ for d.c.) but restricts the a.c. components to flow. Now, if any component of a.c. i.e., pulsating d.c. is still present then it is effectively passed by the capacitor in the shunt ($\therefore X_C << R_L$) because current takes minimum resistance path or in other words we can say that the filter elements are chosen so that the a.c. variation appears across the inductor L instead of load and capacitor C making the filter more effective.

In general, load is high such that,

$$R_L >> 2\omega L >> \frac{1}{2\omega C}$$

Where,

R_L = Load Resistance

ω = Frequency

C = Capacitor

L = Inductor

In actual practice we can't get pure d.c. across the load because some ripples have not yet been filtered, however, in most of the applications it is practically avoidable. The rectified and filtered output for a L-section filter is given in following figure.

From the below figure, we can see that the filtered output starts traversing from zero instant and readily goes towards negative side i.e., below the constant d.c. level output. The reason behind this phenomenon in that the shunt capacitor when charges, takes the voltage from the d.c. level and the filtered output is diminished. When capacitor discharges during its operation, it adds up the current to the load R_L and voltage level gets slightly increased.

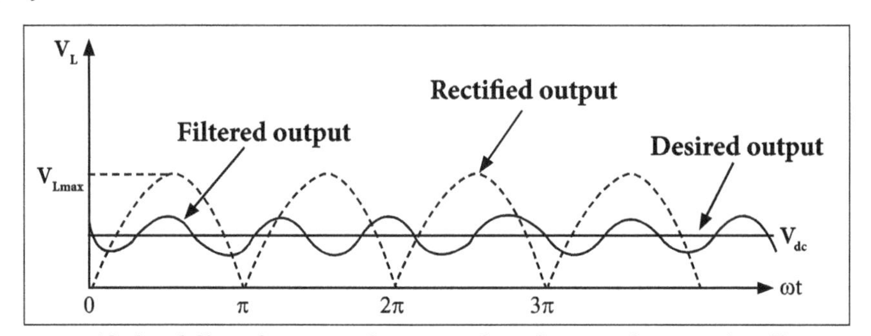

Rectified and filtered output voltage waveform for FWR for L-section filter.

The expression for ripple factor for choke input filter can be expressed as,

$$\text{Ripple factor, } \gamma = \frac{V_{ac\ rms}}{V_{dc}} = \frac{\sqrt{2}}{3}\frac{X_C}{X_L}$$

$$\gamma = \frac{\sqrt{2}}{3}\frac{1}{2\omega C}\cdot\frac{1}{2\omega L} = \frac{1}{6\sqrt{2}\,\omega^2 LC}$$

OR,

$$\gamma = \frac{1}{6\sqrt{2}\,\omega^2 LC}$$

Π-section Filter

In this class of configuration the output from a rectifier is first fed across the capacitor and named as capacitor input filter. The circuit diagram for capacitor input filter is given in the below figure:

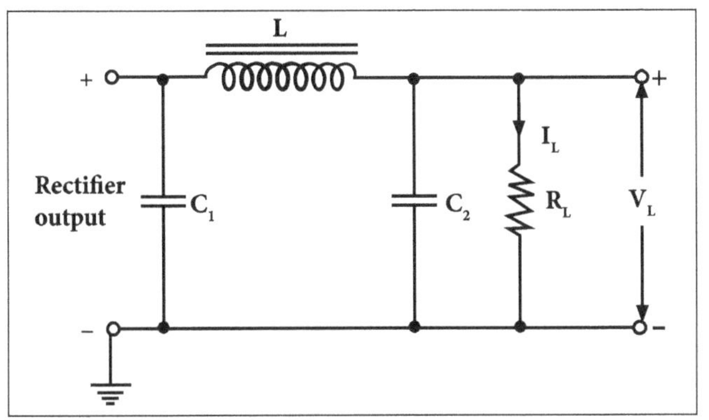

Circuit diagram of a π-Filter.

Basically a π-filter is a combination of two capacitors and one inductor.

It consists of two stages:

- Capacitor shunt filter.
- Choke input filter.

The above figure consists of a shunt capacitor C_1 at the input terminal. An induction L in them connected to C_1 in series. This inductor L is again shunted by another capacitor C_2 and finally load is connected across the C_2. This type of arrangement of filter circuit is often referred to as n-filter just because of ith shape which resembles to the symbol π (Pie).

π- Filter is result of need of higher output voltages at low magnitude loads. To form of π-filter, a input capacitor is shunted to a choke input filter.

The most important feature of this filter is that it can be used for a half wave rectifier circuit because rectifier output is directly connected to the capacitor.

In general, both the capacitors are confined in a single metal container and the metal acts as a common ground for both the capacitors. It is notable here that in filter circuits we employ electrolytic capacitors. The main filtering action is performed by the input capacitor C_1 in this case.

The ripple factor of π-filter is product of the ripple factor of capacitor shunt filter stage and ripple factor of choke input filter stage. If compared with section filter, π-filter have a higher output voltage but the voltage regulation is poor. If any ripples are still present in the output C_1, series inductor L and shunt capacitor C_2 again smoothen the output and a desirable destabilized d.c. output gets associated with the load. The main disadvantage of π-filter is that the output voltage of π-filter falls off rapidly with the increase in load.

Comparison between L-section and π-Filter:

- In a π-filter ripples are less in comparison to shunt capacitor or L-section filter.
- A π-filter requires an inductor of relatively low magnitude than that used in L-section filter.
- Voltage regulation in case of π-filter is poor, so π filters are generally employed with fixed loads, while §ion filters suit better for varying loads.
- PIV is larger in case of a n-filter than in case of L-section filter.
- Current and therefore voltage-regulation for this filter circuit is usually poor. The rectified output and filtered-output in case of capacitor input filter is as given under here.

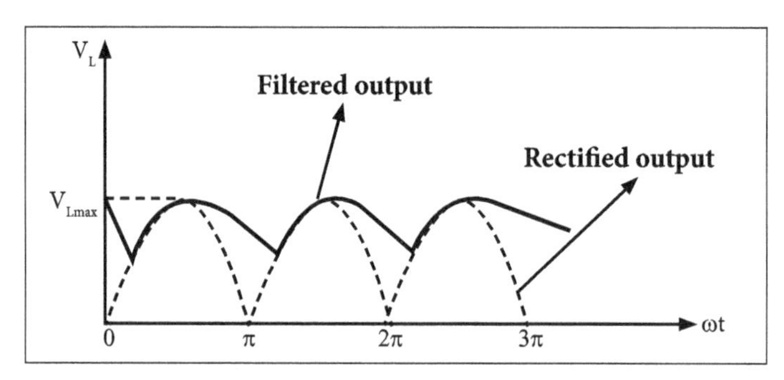

Rectified and filtered output voltage waveforms for a full wave rectifier with capacitor input-filter:

- π- Filter provides better d.c. output voltage than that obtained in the case of L-section filter under similar input conditions.

Expression for ripple factor is given under here:

Ripple factors,

$$\gamma = \frac{V_{ac\ rms}}{V_{dc}} = \frac{\sqrt{2}\, I_{dc}\, X_{C1}\, X_{C2}}{V_{dc}\, X_L} \qquad (\because V_{dc} = I_{dc} . R_L)$$

$$= \frac{\sqrt{2} I_{dc} X_{C1} X_{C2}}{I_{dc} R_L X_L} = \frac{\sqrt{2} X_{C1} X_{C2}}{R_L X_L}$$

or,

$$\gamma = \frac{\sqrt{2}}{R_L} \cdot \frac{1}{\omega C_1} \frac{1}{\omega C_2} \cdot \frac{1}{\omega L} = \frac{\sqrt{2}}{\omega^3 C_1 C_2 L R_L}$$

Advantages of π-filter:

- Reduction in the ripples.
- Simple circuitry.
- Increase in the average load voltage.

Disadvantages of π-filter:

- Diodes-handle large peak currents.
- Ripple factor is dependent on the load.
- Regulation is relatively poor.

3.3.3 Comparison of Various Filter Circuits in Terms of Ripple Factors

The below table shows the comparison of various types of filters when used with the full-wave circuits. In all these filters, the resistances of diodes, transformer and filter elements are neglected and a 60 Hz power line is assumed.

Comparison of Various Types of Filters

	Type of Filter				
	None	L	C	L- section	π – section
V_{dc} at no load	0.636 V_m	0.636 V_m	V_m	V_m	V_m
V_{dc} at load I_{dc}	0.636 V_m	0.636 V_m	$V_m - \frac{4170 I_{dc}}{C}$	0.636 V_m	$V_m - \frac{4170 I_{dc}}{C}$
Ripple factor	0.48	$\frac{R_L}{16000 L}$	$\frac{2410}{CR_L}$	$\frac{0.83}{LC}$	$\frac{3330}{LC_1 C_2 R_L}$
Peak inverse voltage PIV	$2V_m$	$2V_m$	$2V_m$	$2V_m$	$2V_m$

Problems

1. Let us design a filter for full wave circuit with LC filter to provide an output voltage of 2.5 V with a load current of 100 mA and its ripple is limited to 3%.

Solution:

Design of filter means to calculate the value of L and C used in LC filter.

Given that,

$$V_L = 25\,V \text{ and } I_L = 200\,mA$$

So, load resistance,

$$R_L = \frac{V_L}{I_L} = \frac{25\,V}{200\,mA} = 125\ \Omega$$

We know that ripple factor for a LC filter is given by

$$\gamma = \frac{1}{6\sqrt{2}\omega^2 LC}$$

Given that,

$$\gamma = 3\% = .03$$

$$.03 = \frac{1}{6\sqrt{2}\times(2\pi f)^2 LC}$$

$$.03 = \frac{1}{6\sqrt{2}\times 4\pi^2 \times 50^2 LC} \qquad (\because f = 50\,Hz)$$

$$LC = \frac{1}{.03\times 836614.8} = \frac{1}{25098.4} \qquad ...(1)$$

We know that,

$$L = \frac{R_L}{3\omega} = \frac{R_L}{3.(2\pi f)}$$

or

$$L = \frac{125}{3\times 2\times 3.14\times 50}$$

$$L = .1326\ H$$

or

$$C = \frac{1}{.1326 \times 25098.4} = \frac{1}{3330.46} = 300.25\mu F.$$

2. (a) Let us find the output voltage, current and ripple for the circuit shown in the below figure.

(b) Let us find the maximum value of R_L

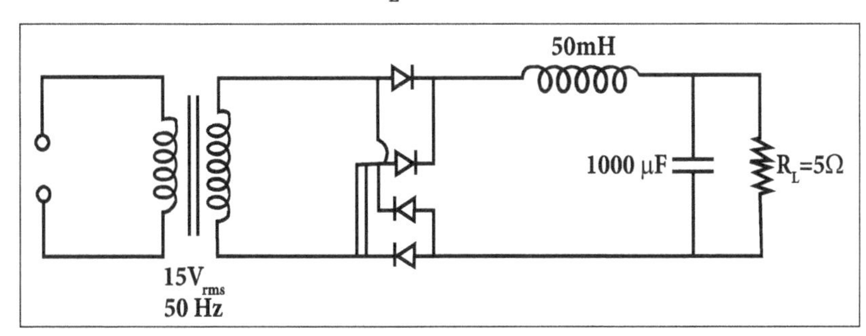

Solution:

(a) $V_{dc} = \frac{2V_m}{\pi} = \frac{2\sqrt{2}\,V_m}{\pi} = \frac{2\sqrt{2} \times 15}{\pi} = 13.5$

$$I_{dc} = \frac{V_{dc}}{R_L} = \frac{13.5}{5} = 2.7\,A$$

$$\gamma = \frac{1}{6\sqrt{2}\,\omega^2 LC} = \frac{1}{6\sqrt{2}(2\pi f)^2 LC}$$

$$= \frac{1}{6\sqrt{2}(314)^2 .50 \times 10^{-3} \times 1000 \times 10^{-6}}$$

The circuit of LC filter is shown in the above figure.

$$\gamma = \frac{1}{41.83} = .024$$

(b) Peak a.c current through inductor

$$I_m \cong \frac{V_m}{X_L}$$

$$\cong \frac{15\sqrt{2}}{2\pi fL} \cong \frac{15\sqrt{2}}{2\times 3.14\times 50\times 50\times 10^{-3}}$$

$$\cong 1.35\,A.$$

$\therefore$ Minimum d.c. Current to avoid spiking (I_{Lmin}) = 1.35A

$$R_{Lmax} = \frac{V_{dc}}{I_{L_{min}}} = \frac{13.5}{1.35} = 10\Omega$$

3. A single phase full-wave rectifier makes use of π-section filter with two 10 μF capacitors and a choke of 104. The secondary voltage is 280 V_{rms} with respect to center tap output. If the load current is 100 mA, let us determine the percentage ripple. Given that V_{dc} = 346 V.

Assume supply frequency of 50 Hz.

Solution:

Given:

$$V_{rms} = 280\text{ V}$$

$$V_{Lmax} = V_m = \sqrt{2}\,V_{rms} = \sqrt{2}\times 280 = 396\,V$$

Load current,

$$I_{dc} = 100\,mA$$

$$C_1 = C_2 = 10\ \mu F$$

$$f = 50\,Hz$$

Ripple factor in case of n-section filter is given by the relation

$$\gamma = \frac{\sqrt{2}}{8\omega^3 C_1 C_2 LR_L} \qquad \left\{ \therefore R_L = \frac{V_{dc}}{I_{dc}} = \frac{346}{100\times 10^{-3}} = 3460 \right\}$$

or

$$\gamma = \frac{\sqrt{2}}{8(2\pi\times 50)^3 \times 10\times 10^{-6}\times 10\times 10^{-6}\times 104\times 3460}$$

or

$$\gamma = .00165$$

Percentage ripple = .165%

3.4 Zener Diode, Types of Regulators and Over Load Voltage Protection

Zener Diode

A Zener diode under reverse bias breakdown condition is used to regulate the voltage across a load irrespective of the supply voltage or load current variations. A simple Zener voltage regulator circuit is shown in the below figure. The Zener diode is selected based on Vz which is the voltage desired across the load.

In Zener diode, under the reverse bias condition, the voltage across it remains constant even though current through it changes by a large extent.

Under normal conditions, the input current is given as,

$$I_i = I_L + I_z$$

Where,

I_i flows through the resistor R.

Input voltage (V_i) is written as,

$$V_i = I_i R + V_z = (I_L + I_z)R + V_z$$

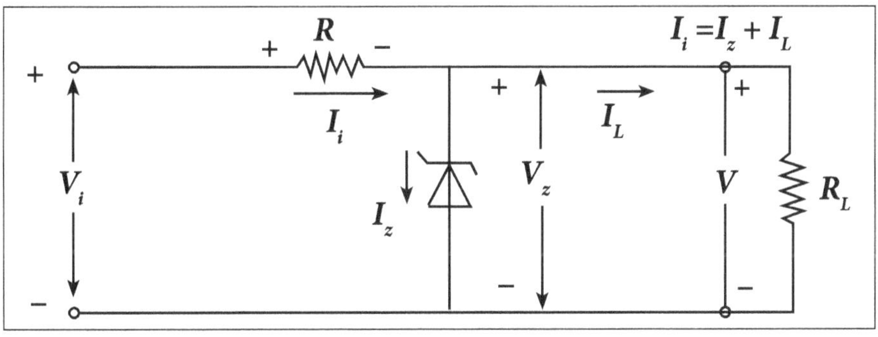

Zener voltage regulator.

The input voltage (V_i) increases when the voltage across Zener diode remains constant and the drop across resistor R will increase with a corresponding increase in $I_L + I_Z$. Since V_z is constant, the voltage across the load remains constant and IL will be constant. Therefore, an increase in $I_L + I_z$ will result in an increase in I_z which will not be altered when the voltage is across the load.

The reverse voltage applied to the Zener diode never exceeds PIV of the diode and at the same time, the applied input voltage is greater than the breakdown voltage of the Zener diode for its operation. The Zener diodes is used as stand-alone regulator circuits and also as reference voltage sources.

3.4.1 Types of Regulators: Series and Shunt Voltage Regulators

Depending on the control element connected in the regulator circuit, the regulators are classified into two types as follows:

- Series voltage regulator.
- Shunt voltage regulator.

Series Voltage Regulator

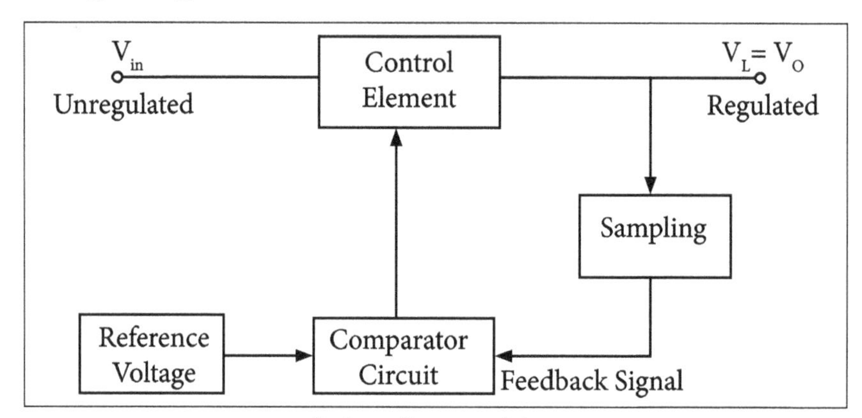

Series voltage regulator.

Here, the control element is connected in series with the load. The unregulated DC voltage is input to the circuit. Control element controls the amount of the input voltage.

The sampling circuit provides the necessary feedback signal. The comparator circuit compares the feedback with the reference voltage to generate the appropriate control signal.

When the load voltage increases, the comparator generates a control signal based on the feedback information. Control signal causes the control element to decrease the constant output voltage.

Shunt Voltage Regulator

The below figure illustrates the equivalent resistive network of a shunt regulator. The location of the regulating device is shown in relation to the load impedance (R_L). Regulation is accomplished in the shunt regulator by division of current between the regulating device and load impedance. When the load demands more current, less current is diverted through the regulator (R_V), allowing more current flow through R_L.

Efficiency of shunt regulators is low under light load conditions (i.e., when current flow through the load is minimal) because the majority of current is being drawn by the regulator. Under full load conditions regulator efficiency is high because minimum current is drawn by the regulator. A significant advantage of a shunt regulator, as opposed to a series regulator, is that it will not be overloaded under load short circuit conditions.

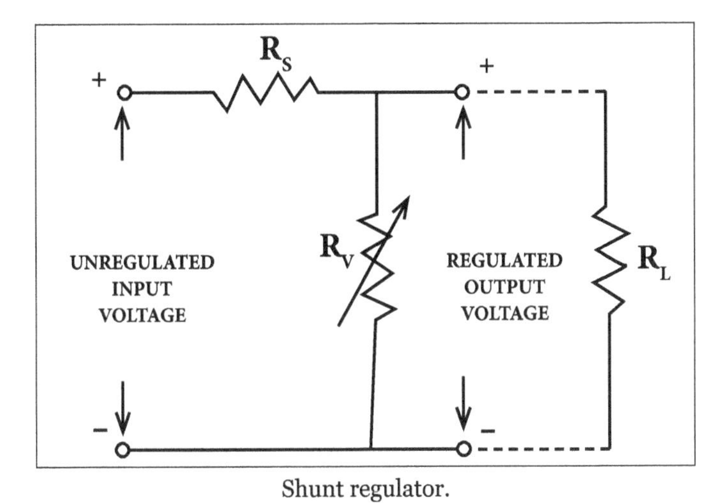

Shunt regulator.

Solid State Shunt Voltage Regulator

The below figure depicts a solid state shunt regulator. Q_1, in parallel with the load impedance, is an NPN transistor with the collector connected to the positive side of the voltage supply and the emitter connected to the negative side through CR_1, a zener diode. CR_1, when reverse biased to its breakdown voltage by R_2, maintains a constant reference voltage at the emitter of Q_1. The base voltage of Q_1 is determined by the setting of potentiometer R_1. This voltage is adjusted so that the base is positive with respect to the fixed emitter potential, forward biasing Q_1, and causing it to conduct. The setting of determines the amount of current through Q_1.

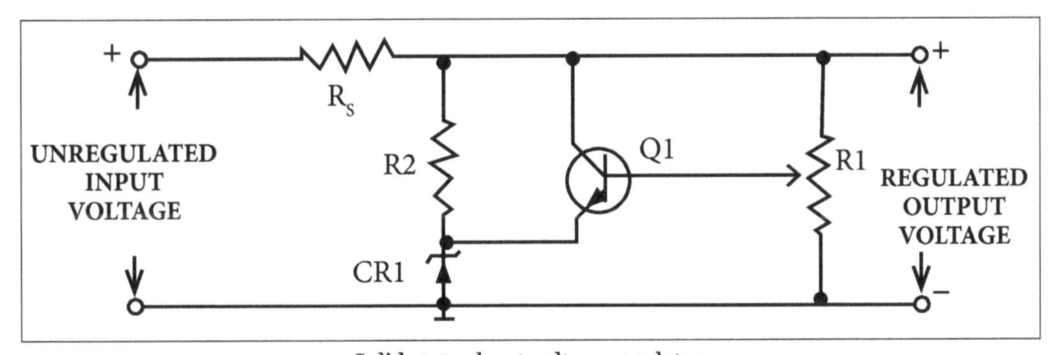

Solid state shunt voltage regulator.

The regulated output voltage is equal to the available supply voltage minus the drop across R_S, the series dropping resistor. The voltage drop across R_S is controlled by the amounts of current drawn by Q_1. Thus, the setting of R_1 will determine the value of the regulated output voltage.

Regulation is accomplished in the following manner. If, for any reason the output voltage increases, the drop across R_1 will increase. This will cause an increased positive potential at the base of Q_1. Since Q_1's emitter is at a fixed potential, due to CR_1, the more positive base will cause Q_1 to conduct more. Current flow through R_S will increase due to the increased transistor current. This will cause an increased voltage drop across R_S, reducing the output voltage to the desired level.

For a decrease in output voltage, the regulation process is reversed. The decreased drop across R_1 will decrease the forward bias of Q_1, causing the transistor to conduct less. Current through R_S will decrease, causing voltage across R_S to decrease, increasing the output voltage.

The regulation process is essentially the same as that occurring in a shunt zener diode regulator. However, the current handling capabilities are greatly increased.

Advantages of shunt voltage regulator:

- It is used in low output voltage switching power supplies.
- It is used in current source and sinks circuits.
- It is used in error amplifiers.
- It is used in adjustable voltage or current linear and switching power supplies.
- It is used for monitoring voltage.
- It is used in analog and digital circuits that require precision references.
- It is used in precision current limiters.

3.4.2 Over Load Voltage Protection

An electrical power system suffers from an abnormal over voltage which is due to various reasons such as sudden interruption of heavy load, lightening impulses, switching impulses etc.

These over voltage stresses may damage insulation of various equipments and insulators of the power system. Although, all the over voltage stresses are not strong enough to damage insulation of system, these over voltages are avoided to ensure the smooth operation of electrical power system.

These types of destructive and non-destructive abnormal over voltages are eliminated from the system by means of overvoltage protection.

Voltage Surge

The over voltage stresses applied on the power system are transient in nature. Transient voltage or voltage surge is defined as sudden sizing of voltage to a high peak in very short duration.

The voltage surges are transient in nature which exists for a very short duration. The main cause of these voltage surges in power system is due to lightning impulses and switching impulses of the system. But over voltage in the power system is due to insulation failure, arcing ground and resonance etc.,

The voltage surges appear in the electrical power system is due to switching surge, insulation failure, arcing ground and resonance which are not very large in magnitude. These over voltages hardly cross the twice of the normal voltage level.

Proper insulation to the different equipment of power system prevents damage due to the over voltages which is due to very high lightning. If over voltage protection is not provided to the power system, there may be high chance of severe damage.

Switching Impulse or Switching Surge

When a no load transmission line is suddenly switched on, the voltage on the line is twice the normal system voltage. This voltage is transient in nature. When a loaded line is suddenly switched off or interrupted, voltage across the line becomes high which causes over voltage in the system.

During insulation failure, a live conductor is suddenly earthed which causes sudden over voltage in the system. If emf wave produced by alternator is distorted, the trouble of resonance is due to 5th or higher harmonics.

For frequencies of 5th or higher harmonics, a critical situation in the system appears where the inductive reactance of the system is equal to capacitive reactance of the system. As these both reactance cancel each other, the system becomes purely resistive. This phenomenon is called as resonance and at resonance, the system voltage is high.

All these above mentioned reasons create over voltages in the systems which are not very high in magnitude.

But over voltage appear in the system due to lightning impulses which are very high in amplitude and they are highly destructive. The effect of lightning impulse is avoided for over voltage protection of power system.

4

Transistors

4.1 Junction Transistor

P-N-P Bipolar Junction Transistor

For a p - n - p bipolar junction transistor, the n-type semiconductor is sandwiched between the two p-type semiconductors. The diagram of a p - n - p transistor is as shown:

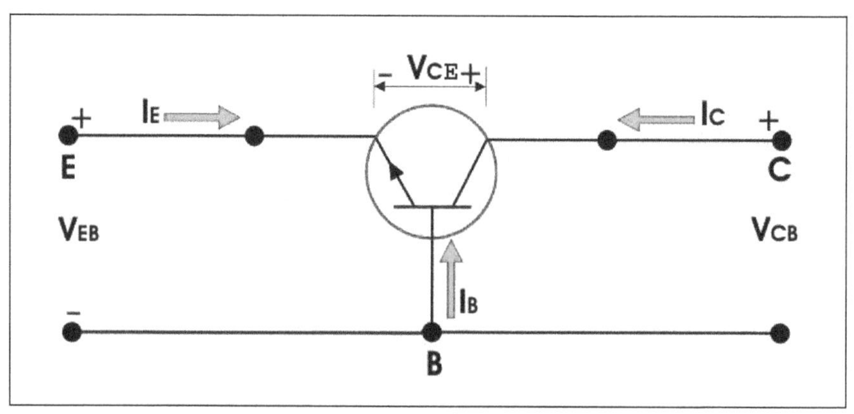

PNP transistor.

For a p - n - p transistors, current enters into the transistor through emitter terminal. Like any BJT, the emitter – base junction is initially forward biased and the collector – base junction is then reverse biased.

Tabulation shows the emitter, base and collector current, as well as the emitter base, collector base and collector emitter voltage for p - n - p transistors.

Transistor type	I_E	I_B	I_C	V_{EB}	V_{CB}	V_{CE}
p - n - p	+	-	-	+	-	-

N-P-N Bipolar Junction Transistor

- In n - p - n bipolar transistor, one p - type semiconductor resides between the two n - type semiconductors. The diagram below is a n - p - n transistor.

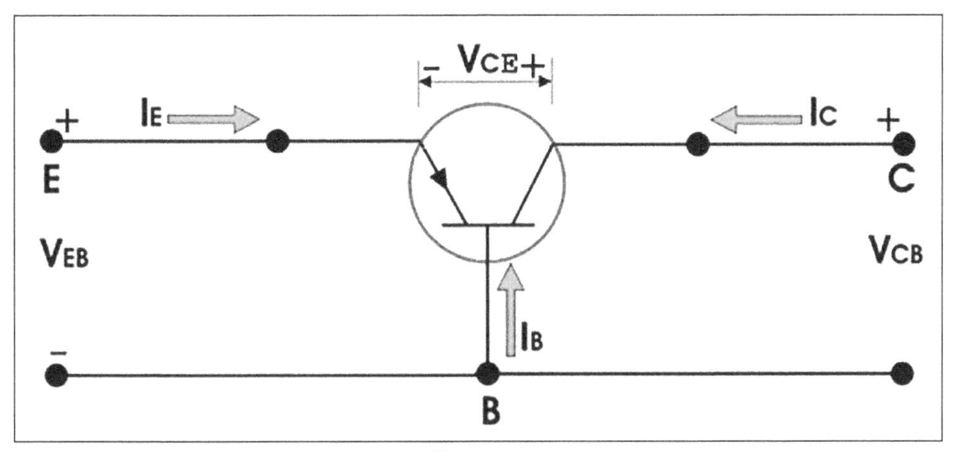

NPN transistor.

Now I_E, I_C is emitter current and collect current respectively and V_{EB} and V_{CB} are the emitter base voltage and the collector base voltage respectively. According to convention, for the emitter, base and collector current I_E, I_B and I_C current goes into the transistor, the sign of the current is taken as positive and if the current goes out from the transistor, then the sign is taken as negative.

We can tabulate different currents and voltages inside the n - p - n transistor.

Transistor type	I_E	I_B	I_C	V_{EB}	V_{CB}	V_{CE}
n - p - n	-	+	+	-	+	+

Early Effect

As the collector voltage V_{cc} is made to increase the reverse bias, the depletion width between collector and base tends to increase with the result that the effective width of the base decreases. This dependency of base-width on collector-to-base voltage is called s the early effect. This decrease in effective base-width has three consequences:

- There is a less chance for recombination within the base region. Hence, α increases with increasing V_{CB}.
- The charge gradient is increased within the base and consequently the current of minority carriers injected across the emitter junction increases.
- For extremely large voltages, the effective base-width may be reduced to zero, causing voltage breakdown in the transistor. This is called as punch through.

For higher values of V_{CB} due to early effect, value α increase. For example, α changes from 0.98 to 0.985. Hence, there is a very small positive slope in the CB output characteristics and thus the output resistance is not zero.

4.1.1 Transistor Current Components

Current Components of BJT Transistor

When BJT transistor is not biased, that is, there is no voltage drop across its junctions and thus, no current flows through it. If Emitter-Base junction is forward biased and Collector-Base junction is reverse biased, the voltage across the device causes electrons from emitter to flow to collector. In this, electrons pass through P type lightly doped base region and some of the electrons recombine with the holes. Thus, collector current is less than that of the emitter current. Emitter current, Base current and Collector current may be related by,

$$\text{Emitter current} = \text{Base current} + \text{Collector current}$$

Mainly these three parameters are used to define BJT transistor performance. Current Amplification factor, Base Transport Factor, Emitter Injection Efficiency parameters shows the performance of the NPN transistor and PNP transistor.

Base Transport Factor

Base Transport Factor is defined as the factor of base current required to transfer emitter current to collector of the BJT transistor. Base transport factor is the ratio of collector current to base current of a BJT transistor i.e., it is the ratio of output current to input current in the common emitter configuration.

$$\beta = I_c / I_b$$

Current Amplification Factor

Current amplification factor in a BJT transistor is defined as the ratio of the output current to the corresponding input current. In a common base configuration, the current amplification factor is ratio of the collector current to the emitter current.

$$\alpha = I_c / I_e$$

Emitter Injection Efficiency

Emitter injection efficiency in a BJT transistor refers to the efficiency of majority carrier injection from emitter. It is ratio of current due to emitter majority carriers to the total emitter current. It defines the injection capability of an emitter. The heavily doped region will have high injection factor.

4.1.2 Characteristics of Transistor: CE, CB and CC Configurations

Ce-Configuration

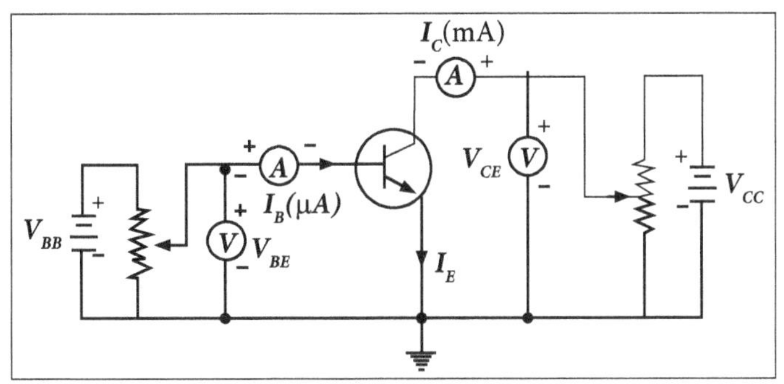

Circuit for CE static characteristics.

Input Characteristics of a Transistor in Common - Emitter Configuration

These characteristics may be obtained by using the circuit arrangement shown in the figure above. First of all, we adjust the collector-to-emitter voltage (V_{CE}) to 1 volt. Then, we can increase the base-to-emitter voltage (V_{BE}) in small suitable steps and record the corresponding values of base current (I_B) at each step.

If we plot a graph with base-to-emitter voltage (V_{BE}) along the vertical axis, we shall obtain a curve marked V_{BE} = 1 V as shown in figure. Similar procedure may be used to obtain characteristics at different values of collector-to-emitter voltage, i.e., V_{CE} = 2, 10 and 20V. The input characteristics gives us the information about the following important point:

- There exists a threshold or knee voltage (V_K) below which the base current is very small. The value of the knee voltage is 0.5 V for silicon and it is 0.1 V for germanium transistors.
- Beyond the knee, the base current (I_B) increases with the increase in base-to-emitter voltage (V_{BE}) for a constant collector-to-emitter voltage (V_{CE}). However, it may be noted that the value of base current does not increase rapidly as that of the input characteristics of a common base transistor which means that input resistance of a transistor in common-emitter configuration is higher as compared to the common base configuration.
- Since the collector-to-emitter voltage (V_{CE}) is increased above 1 $_{V'}$ the curves shift downwards. It occurs because of the fact that as V_{CE} is increased, the depletion width in the base-region increases. This reduces the effective base width which in turn reduces the base current.
- Input characteristics may be used to determine the value of common-emitter transistor a.c. input resistance (R). Its value is given by the ratio of change in

base to emitter voltage (ΔV_{BE}) to the resulting change in base current (I_B) at a constant collector to emitter voltage (V_{CE}). Mathematically the ac input resistance is given by,

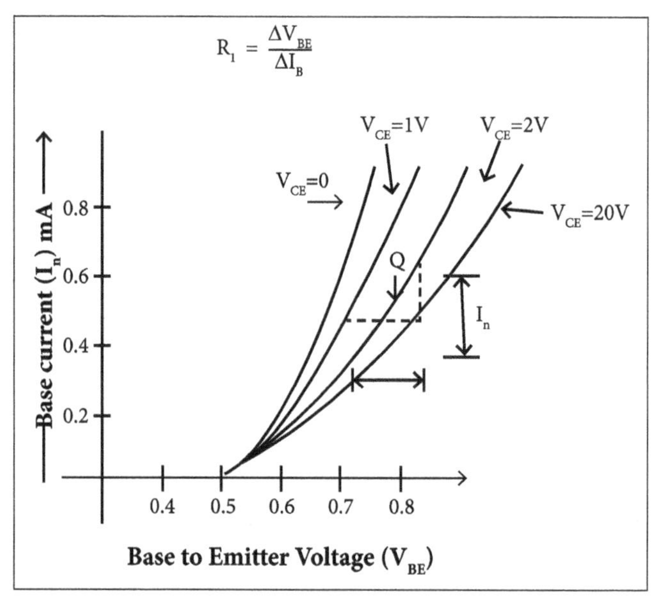

Input characteristics curve.

It may be noted that the input characteristic is not linear in the lower region of the curve. Therefore, the input resistance varies with the location of the operating point. The value of ac input resistance ranges from 600 Ω to 4100 Ω.

These characteristics may be obtained by using the circuit shown above. To begin with, increase the base current (I_B) to 40mA value. Then increase the collector-to-emitter voltage (V_{CE}) in a number of steps and fire cord the corresponding values of collector current (I_c) at each step.

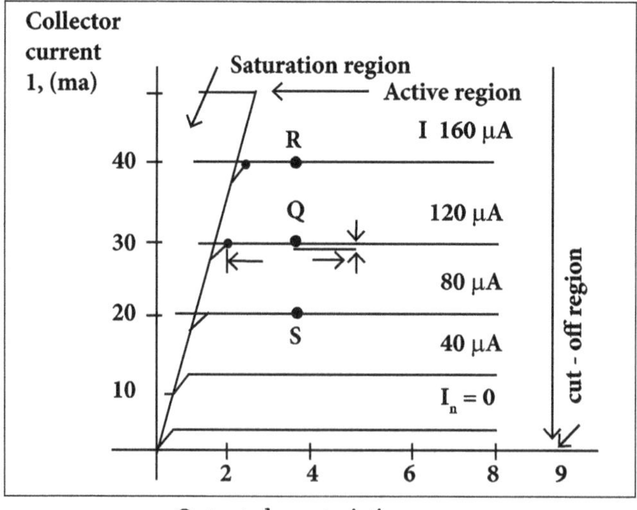

Output characteristic curve.

If we plot a graph with collector-to-emitter voltage (V_{CE}) along the horizontal axis and collector current (I_c) along the vertical axis, we shall obtain a curve marked I_B = 40mA as shown in figure. A similar procedure may be used to obtain characteristics at the I_B = 80mA, 120mA, 160mA and so on. The output characteristics give us the information about the following important points.

- The output characteristics may be divided into three important regions namely saturation region, active region and cut-off region while the active region is the region between the saturation and cutoff region.
- As the collector to emitter voltage (V_{CE}) is increased above zero, the collector current (I_c) increases rapidly to a saturation value depending upon the value of base current. It may be noted that the collector current (I_c) reaches to a saturation value when V_{CE} is about 1 V.
- When collector-to-emitter voltage (V_{CE}) is increased further, the collector current slightly increases. This increase in collector current is due to the fact that increased value of collector to emitter voltage (V_{CE}) reduces the base current and hence the collector current increases. This phenomenon is called an early effect.
- When the base current is zero, a small collector current exists. This is called leakage current. However for all practical purposes, the collector current (I_c) is zero when the base current (I_B) is zero, under this condition the transistor is said to be cut-off.
- The characteristic may be used to determine the common emitter transistor ac output resistance. Its value at any given operating point Q is given by the ratio of change in collector-to-emitter voltage (ΔV_{CE}) to the resulting change in collector current (ΔI_C) for a constant base current. Mathematically, the ac output resistance is expressed as,

  $$R_o = \frac{\Delta V_{CE}}{\Delta I_C}$$

  The common emitter output resistance of a transistor ranges from 10 kΩto 50 kΩ.
- The characteristic may be used to determine the small signal common emitter current gain or ac beta (βo) of a transistor. This can be done by selecting two points R and S on the characteristics and note down the corresponding value of ΔI_C and ΔI_B. Thus if, $\Delta I_C = 40 - 20 = 20$ mA $= 20 \times 10^{-3}$ A and $\Delta I_B = 160 - 80 = 80$ mA $= 80 \times 10^{-6}$ A then small signal common emitter current gain,

  $$\beta_o = \frac{\Delta I_C}{\Delta I_B} = \frac{20 \times 10^{-3}}{80 \times 10^{-6}} = 250.$$

Cb-Configuration

Input Characteristics

To determine the input characteristics, the collector to emitter voltage is kept constant at zero volts and the base current is increased from zero in equal steps by increasing V_{BE}.

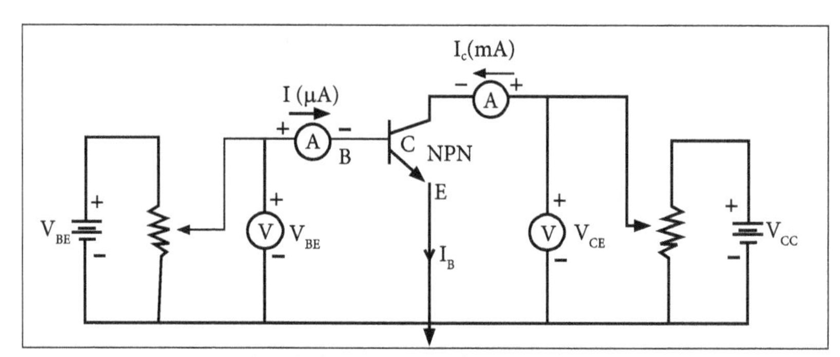

Circuit for CB static characteristics.

The value of V_{BE} is noted for each setting of I_B. This procedure is repeated for higher fixed values of V_{CE} and the curves of I_B vs V_{BE} are drawn.

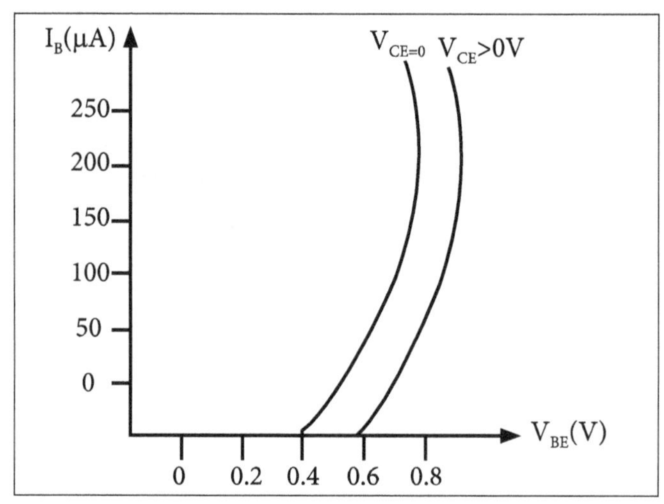

Input characteristic curve.

When V_{CE} = 0, the EB junction is forward biased and the junction behaves as a forward biased diode. Hence, the input characteristic for V_{CE} = 0 is similar to that of a forward biased diode.

When V_{CE} is increased, the width of the depletion region at the reverse biased collector-base junction will increase. Hence, the effective width of the base will decrease. This effect causes a decrease in the base current I_B. Hence to get the same value of I_B as that for V_{CE} = 0, V_{BE} should be increased. Therefore, the curve shifts to the right as V_{CE} increases.

Output Characteristics

To determine the output characteristics, the base current I_B is kept constant at a suitable value by adjusting V_{BE}. The magnitude of V_{CE} is increased in suitable equal steps from zero and the value of I_c is noted for each value of V_{CE}. Now the curves of I_c vs V_{CE} are plotted for different constant values of I_B.

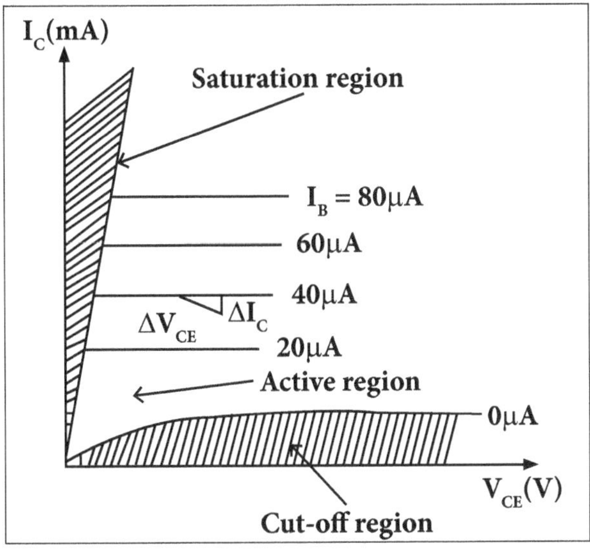

Output characteristics curve.

The output characteristics have three regions namely:

- Saturation region.
- Cut-off region.
- Active region.

Saturation Region

In this region, both the junctions are forward biased and an increase in the base current does not cause a corresponding large change in I_c. The ratio of V_{CE} (sat) to I_c in this region is called Saturation resistance.

Cut-off Region

In this region, both the junctions are reverse biased. Transistor is in OFF state. Hence, the collector current becomes almost zero and the collector voltage almost equals V_{CC}. The transistor is virtually an open circuit between collector and emitter.

Active Region

The central region where the curves are uniform in spacing is called as the active region.

In this region, EB junction is forward biased and the collector base junction is reverse biased. If the transistor is to be used as a linear amplifier, it should be operated in active region.

Problem

In a common base connection if the emitter current is 1 mA, I_{CBO}=50 μ A, α = 0.92. Total collector current for this circuit can be determined as follows.

Solution:

$$I_E = I mA, I_{CBO} = 50\mu A, \alpha = 0.92, I_C = ?$$

$$\alpha = \frac{I_C}{I_E}.$$

$$I_C = \alpha I_E = (0.92)(1 mA)$$

$$= 0.92 mA.$$

Cc-Configuration

Common collector configuration circuit is shown in the figure below. Here collector is grounded and it is used as the common terminal for both input and output. It is also called as grounded collector configuration. Base is used as a input terminal whereas emitter is the output terminal.

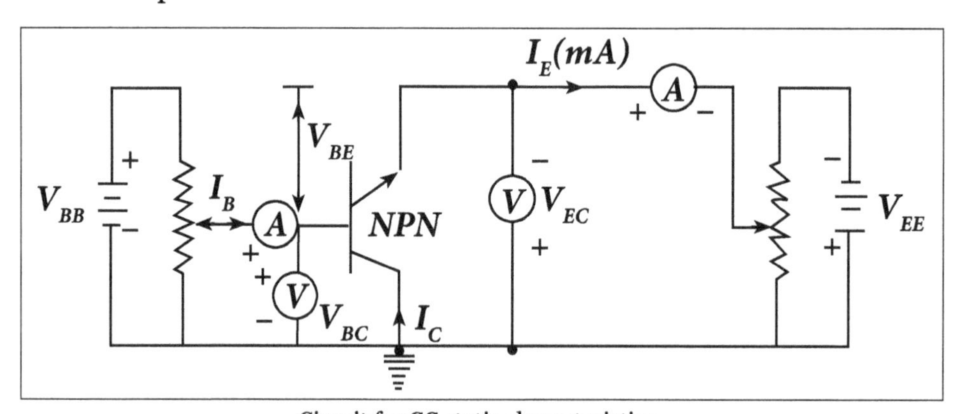

Circuit for CC static characteristics.

Input Characteristics

It is defined as the characteristic curve drawn between input voltages to input current whereas output voltage is constant.

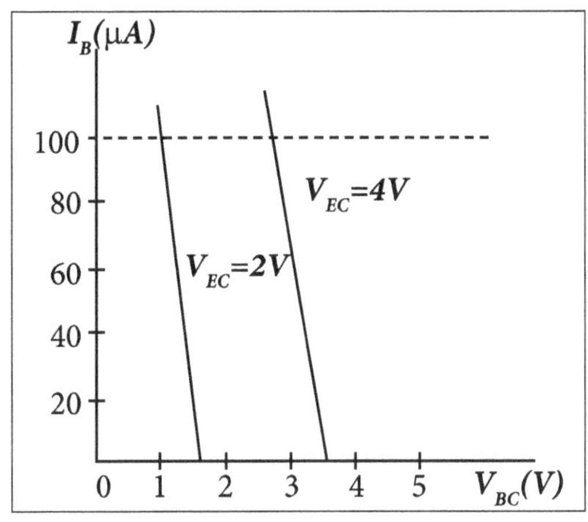

Input Characteristic curve.

To determine input characteristics, the emitter base voltage V_{EB} is kept constant at zero and base current I_B is increased from zero by increasing V_{BC}. This is repeated for higher fixed values of V_{CE}.

A curve is drawn between base current and base emitter voltage at constant collector base voltage as shown in the above figure.

Output Characteristics

It is defined as the characteristic curve drawn between output voltages to output current whereas input current is constant.

To determine the output characteristics, the base current I_B is kept constant at zero and emitter current I_E is increased from zero by increasing V_{EC}. This is repeated for higher fixed values of I_B.

From the characteristic it is seen that for a constant value of I_B, I_E is independent of V_{EB} and the curves are parallel to the axis of V_{EC}.

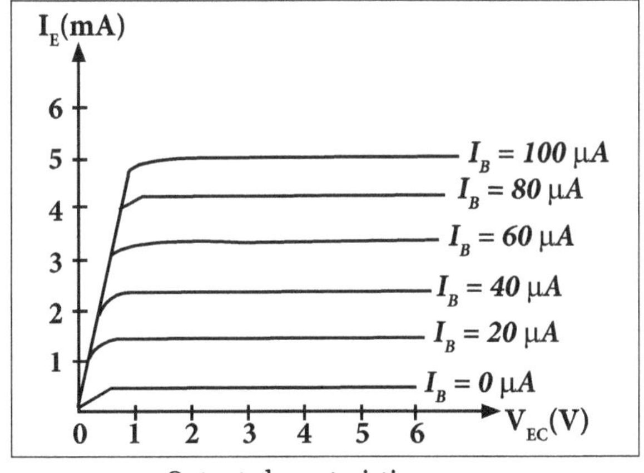

Output characteristic curve.

Performance of a Transistor in Various Configurations

S. No.	Property	CB	CE	CC
1.	Input resistance	low	moderate	high
2.	Output resistance	high	moderate	low
3.	Voltage gain	high	high	≈1
4.	Current gain	≈1	high	high
5.	Phase shift between input voltage end output voltage	0° to 360°	180°	0° or 360°

Relationship between α and β

We know that,

$$\Delta I_E = \Delta I_C + \Delta I_B$$

By definition,

$$\Delta I_C = \alpha\, \Delta I_E$$

Therefore,

$$\Delta I_a = \Delta I_E (1-\alpha)$$

Dividing both sides by ΔI_C, we get,

$$\frac{\Delta I_B}{\Delta I_C} = \frac{\Delta I_E}{\Delta I_C}(1-\alpha)$$

Therefore, $\frac{1}{\beta} = \frac{1}{\alpha}(1-\alpha) \Rightarrow \beta = \frac{\alpha}{1-\alpha}$

Rearranging, we also get $\alpha = \frac{\beta}{1+\beta}$ (or) $\frac{1}{\alpha}\frac{1}{\beta} = 1$

From this relationship it is clear that as α approaches unity, β approaches infinity. The CE configuration is used for almost all transistor applications because of its high current gain, β.

From this relationship it is clear that as α approaches unity, β approaches infinity. The CE configuration is used for almost all transistor applications because of its high current gain, β.

Relation Among α, β and γ

In the CC transistor amplifier circuit, I_B is the input current and I_E is the output current.

We know that, $\gamma = \frac{\Delta I_E}{\Delta I_B}$

Substituting $\Delta I_B = \Delta I_E - \Delta I_C$, we get $\gamma = \frac{\Delta I_E}{\Delta I_E - \Delta I_C}$.

Dividing the numerator and denominator on R.H.S. by ΔI_E, we get

$$\gamma = \frac{\frac{\Delta I_E}{\Delta I_E}}{\frac{\Delta I_E}{\Delta I_E} - \frac{\Delta I_C}{\Delta I_E}} = \frac{1}{1-\alpha}.$$

$$\gamma = \frac{1}{1-\alpha} = (\beta + 1)$$

Problem

The value of a for a transistor is 0.950. Let us calculate the value of β andα if it changes to 100.

Solution:

The relation between α and β is given by,

$$\beta = \frac{\beta}{1-\alpha}$$

$$\beta = \frac{0.950}{1-0.950} \Rightarrow \beta = 19.$$

$$\beta = \frac{\alpha}{1-\alpha} \text{ (or) } \alpha = \frac{\beta}{1+\beta} = \frac{19}{1+19} = 0.95$$

For $\beta = 100$

$$\alpha = \frac{100}{1+100} = \frac{100}{101} = 0.98$$

4.2 Transistor Biasing and Thermal Stabilization

Stability Factor (S)

The extent to which the collector current I_c is stabilized with varying I_{co} can be measured by a stability factor S.

It is defined as the rate of change of collector current I_c with respect to the collector—base leakage current I_{co}, keeping both the current I_B and the current gain β constant.

$$S = \frac{\partial I_C}{\partial I_{CO}} \simeq \frac{dI_C}{dI_{CO}} \simeq \frac{\Delta I_C}{\Delta I_{CO}}, \beta \text{ and } I_B \text{ constant}$$

The collector current for a CE amplifier is given by,

$$I_C = \beta I_B + (\beta + 1) I_{CO}$$

Differentiating the above equation with respect to I_c we get,

$$1 = \beta \frac{dI_B}{dI_C} + (\beta + 1) \frac{dI_{CO}}{dI_C}$$

$$\left(\beta \frac{dI_B}{dI_C}\right) = \frac{(\beta + 1)}{S}$$

$$S = \frac{1 + \beta}{1 - \beta} \left(\frac{dI_B}{dI_C}\right)$$

From this equation, it is clear that this factor S should be as small as possible to have better thermal stability.

Stability Factors S' and S"

The stability factor S' can be defined as the rate of change of I_c with V_{BE} keeping I_{CBO} and B as constant. The stability factor S" can be defined as the rate of change of I_c with respect to 13, keeping I_{co} and β constant,

$$S' = \frac{\partial I_C}{\partial V_{BE}} \simeq \frac{\Delta I_C}{\Delta V_{BE}}$$

The stability factor S" is defined as the rate of change of c with respect to β keeping I_{co} and VBE constant,

$$S'' = \frac{\partial I_C}{\partial \beta} \simeq \frac{\Delta I_C}{\Delta \beta}$$

Bias Compensation

The various biasing circuits considered in the previous sections used some types of negative feedback to stabilize the operation point. Also, diodes, thermistors and sensistors can be used to compensate for variations in current.

Diode Compensation

Figure shows a transistor amplifier with a diode D connected across the base-emitter junction for compensation of change in collector saturation current I_{co}.

The diode is of the same material as the transistor and it is reverse biased by the base-emitter junction voltage V_{BE}, allowing the diode reverse saturation current Io to flow through diode D. The base current $I_B = I - I_o$. As long as temperature is constant, diode D operates as a resistor.

As the temperature increases, I_{co} of the transistor increases. Hence to compensate for this, the base current I_B should be decreased. The increase in temperature will also cause the leakage current Io through D to increase and thereby decreasing the base current I_B . This is the required action to keep I_C constant.

This method of bias compensation does not need a change in I_C to effect the change in I_B, as both I_o and I_{CO} can track almost equally according to the change in temperature.

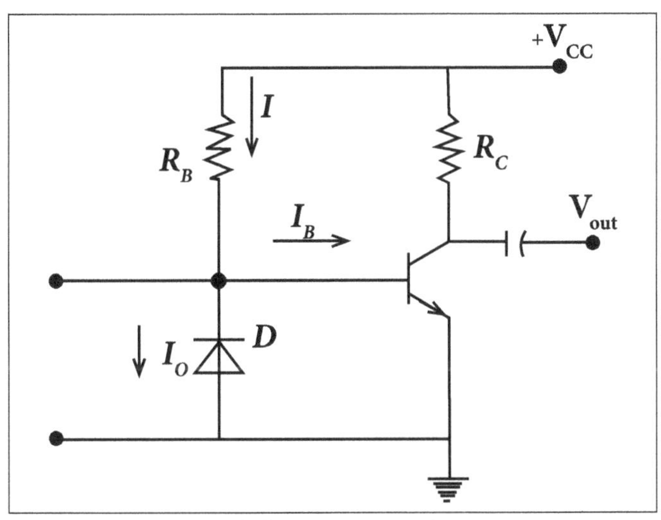

Diode bias Compensation.

Thermistor Compensation

In Figure a thermistor R_T, having a negative temperature coefficient is connected in parallel with R_2.

The resistance of thermistor decreases exponentially with increase of temperature. This increase in temperature will decrease the base voltage V_{BE}, reducing I_B and I_C. Bias stabilization can also be provided by R_E and C_E.

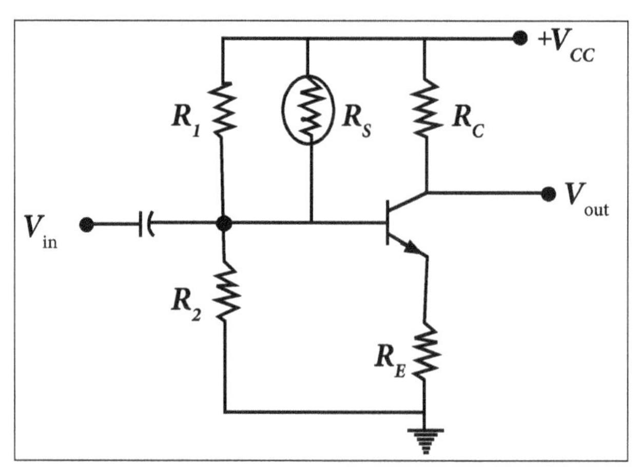

Thermistor bias compensation.

Sensistor Compensation

In Figure a sensistor R_s having a positive temperature coefficient is connected across R_1 (or R_E). R_s increases with temperature. As temperature increases, the equivalent resistance of the parallel combination of R_1 and R_s also increases and hence the base voltage V_{BE} decreases, reducing I_B and I_C this reduced I_C compensates for the increased I_C caused by the increase in I_{CO}, V_{BE} and also for the temperature rise.

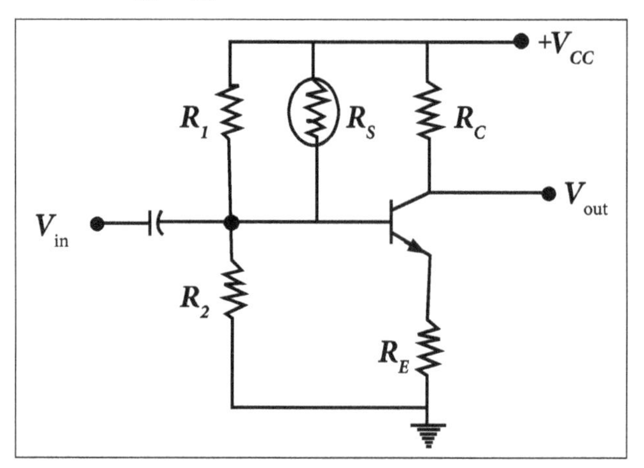

Sensistor bias compensation.

4.2.1 Fixed Bias, Collector to Base Bias and Self-Bias

Biasing Circuits of BJT

To make the Q point stable different biasing circuits are tried. The Q point is also known as operating bias point, is the point on DC load line which represents the DC current through the transistor and voltage across it when no ac signal is applied. The Q point represents DC biasing condition. When the BJT is biased such that the Q point is halfway between cutoff and the saturation than the BJT operates as a CLASS-A amplifier. The three circuits or the biasing arrangements which are practically used are explained.

Fixed Bias or Base Bias

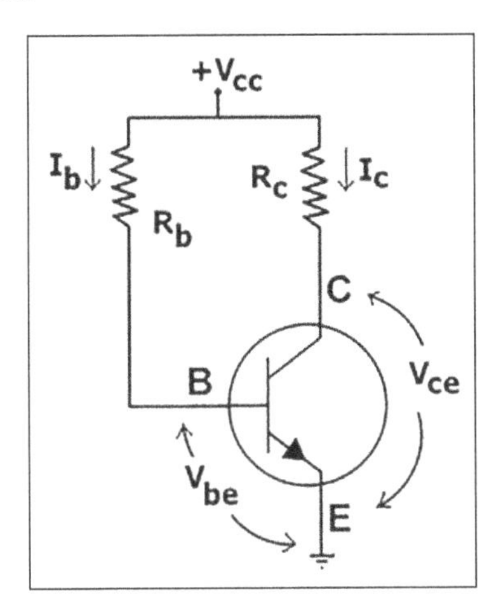

In this condition, the single power source is applied to the collector and base of the transistor using only the two resistors. Applying KVL to the circuit,

$$V_{CC} = I_B R_B + V_{BE}$$

$$\Rightarrow I_B = \frac{(V_{CC} - V_{BE})}{R_B}$$

Thus, by merely changing the value of the resistor the base current can be adjusted to the desired value. And by using the current gain (β) relationship, I_C can also be found out accordingly. Hence the Q point can be adjusted just by changing the value of the resistor connected to the base.

Collector to Base Bias

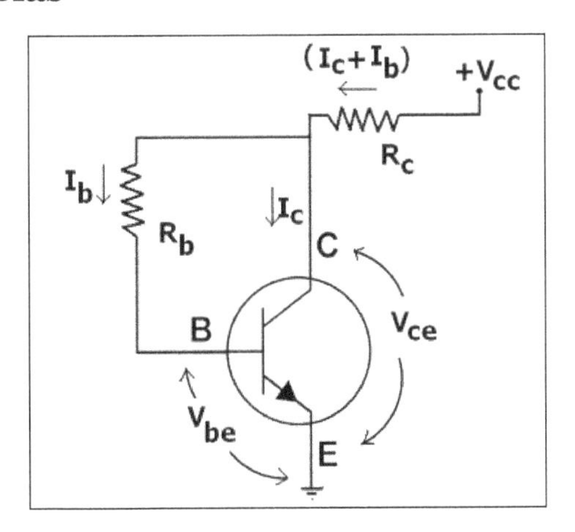

This connection is mostly used to stabilize the operating point against temperature changes. In this type, the base resistor is connected to the collector instead of connecting it to the supply. So any thermal runaway may induce IR drop in the collector resistor. The base current may be derived as,

$$I_b = \frac{V_{CC} - V_{BE}}{R_b + (\beta + 1)R_C}$$

If V_{BE} kept constant and there is an increase in temperature, then the collector current increases. However, a larger collector current causes the voltage drop across the collector resistor to increase, which reduces the voltage across the base resistor. This will reduce the base current, thus resulting less collector current. Because an increase in collector current with temperature is opposed, the operating point is stable.

Self-Bias or Voltage Divider Bias

The circuit diagram for self-bias is shown below. This is the most widely used biasing circuit.

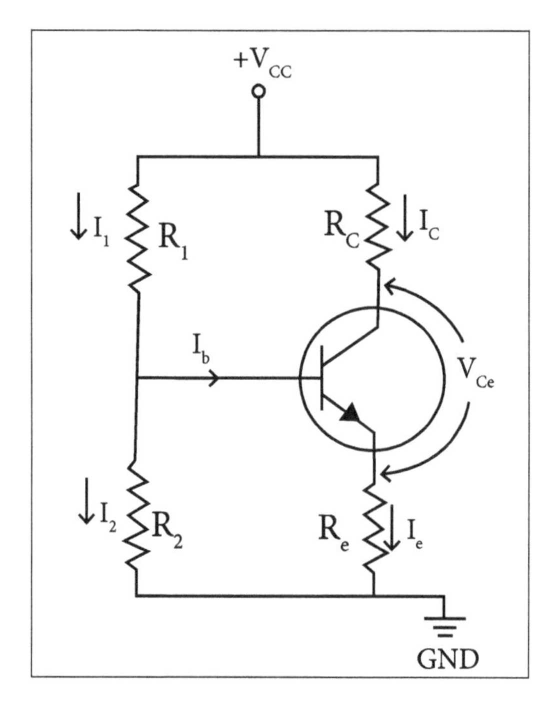

The electrical resistances R_1 and R_2 form a potential divider arrangement to apply a fixed voltage to the base.

Consider only the base circuit, the approximate voltage across the base is,

$$V_B = V_{R2} = \frac{V_{CC} \times R_2}{R_1 + R_2}$$

Consider only the collector circuit; the approximate emitter current will be,

$$I_E = \frac{(V_B - V_{BE})}{R_E}$$

In the above circuit, as the emitter resistor causes AC as well as the DC feedback the AC voltage gain of the amplifier is merely reduced. This can be avoided by connecting a capacitor in parallel with the emitter resistor as shown below:

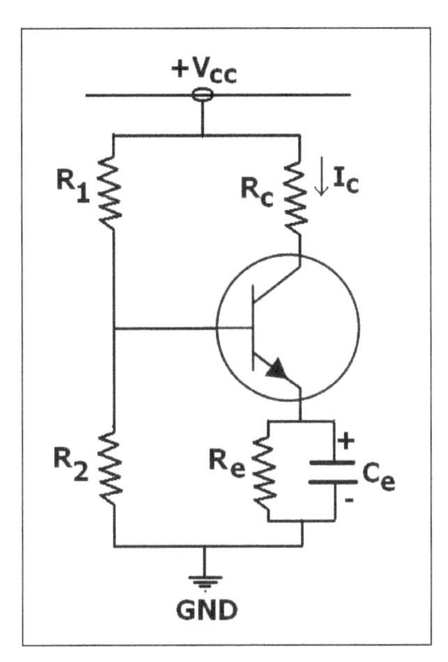

4.2.2 Compensation Against Variation in Base Emitter Voltage and Collector Current

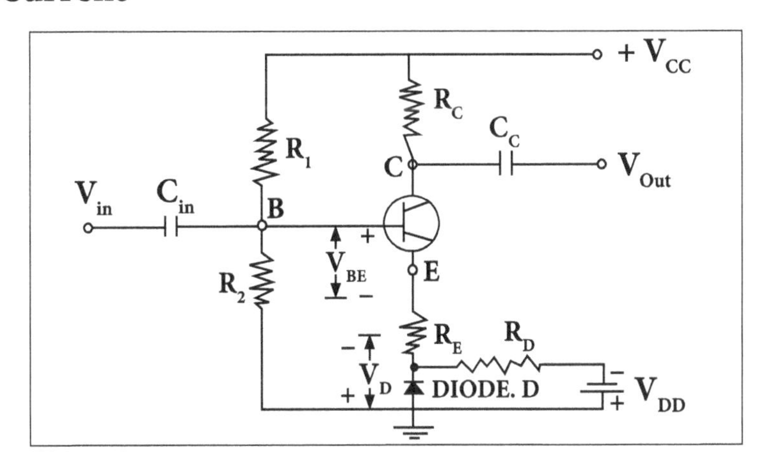

Working of the circuit:

- The circuit utilizes the self-Bias stabilization and the diode compensation using the silicon transistor.

- The diode is kept forward biased by the source VDD and the resistor R_D.
- The diode employed is of the same material and type of the transistor to have the same temperature co-efficient (-2.5mv/°c).

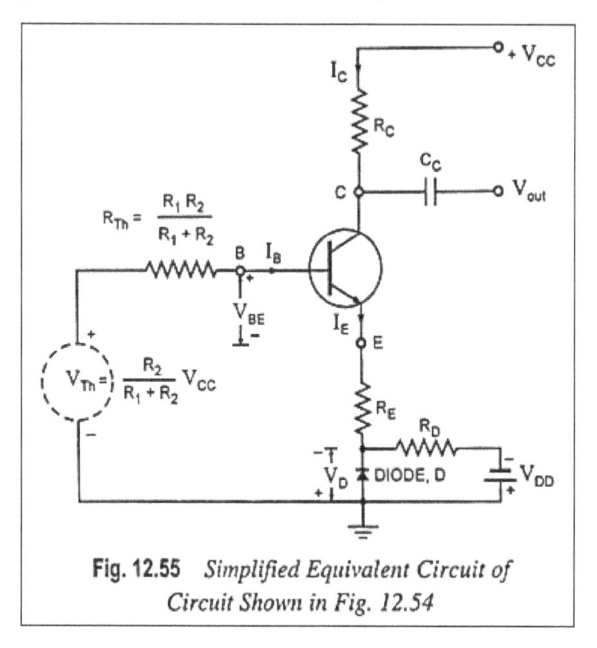

Fig. 12.55 *Simplified Equivalent Circuit of Circuit Shown in Fig. 12.54*

Circuit Analysis

By applying Kirchhoff's voltage law to the base portion,

$$V_{TH} = I_B R_{TH} + V_{BE} + I_E R_E - V_D$$

$$I_E = I_B + I_C$$

$$V_{TH} = I_B R_{TH} + V_{BE} + (I_B + I_C)\ R_E - V_D$$

$$= I_B (R_{TH} + R_E) + I_C R_E + V_{BE} - V_D$$

$$I_C = \beta\ I_B + (1+\beta) I_{CO}$$

$$V_{TH} = \{(I_C - (1+\beta) I_{CO}) / \beta\} (R_{TH} + R_E) + I_C R_E + V_{BE} - V_D$$

$$I_C = \beta\ [V_{TH} - (V_{BE} - V_D)] + (1+\beta)\ I_{CO} (R_{TH} + R_E) / (R_{TH} + (1+\beta) R_E)$$

Variations in V_{BE} and V_D are same due the variation in temperature.

(V_{BE}-V_D) remains unchanged.

Collector current I_C becomes insensitive to variations in V_{BE}

Diode Compensation for Variation in I_{Co}

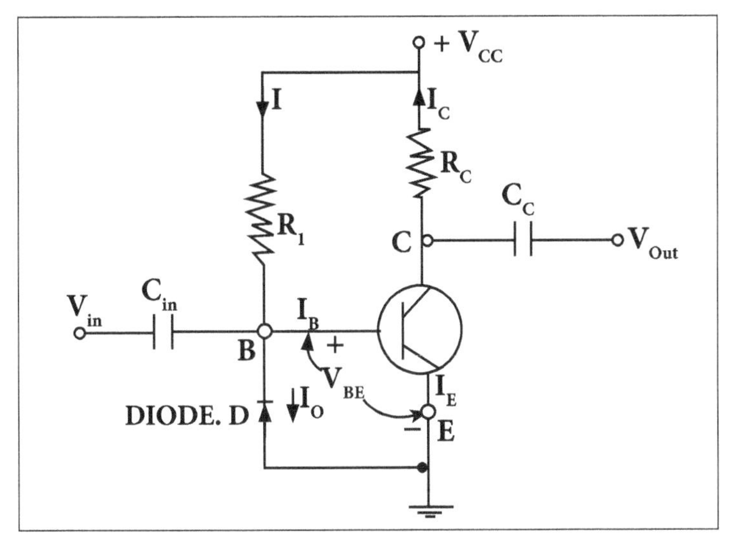

Working of the Circuit:

- This figure shows the circuit using diode compensation for a Germanium transistor.
- In Germanium transistors changes in reverse saturation current I_{Co} with temperature variations causes a change in collector stability.
- The Diode D used in the circuit is of the same material and type as transistor
- The reverse saturation current of the transistor I_{Co} and of the Diode I_o, will increase with increase in the temperature.
- From the circuit diagram

  $I = V_{CC} - V_{BE} / R_1 \cong V_{CC} / R_1$

  $=$ constant

  V_{BE} is very small in comparison with the V_{CC}.
- Since Diode is reverse biased by the V_{BE} the current through diode is reverse saturation current I_o.

  Base current $I_B = I - I_o$

  $I_C = \beta I_B + (1+\beta) I_{Co}$

$$= \beta_I - \beta I_o + (1+\beta)\, I_{Co}$$

4.2.3 Thermal Runaway

Due to unequal current division when current through SCR increases, its temperature also increases which in turn decreases the resistance. Hence further increase in the current takes place and this is a cumulative process. This is known as thermal 'run away' which can damage the device.

To overcome this problem, SCR's will be maintained at the same temperature. This is possible by mounting them on the same heat sink. They must be mounted in symmetrical position as flux. Linkages by the devices will be same. So, the mutual inductance of devices will be the same. This will offer same reactance through every device. Thus reducing the difference in the current level through the devices.

Another way of equalizing current division in an ac circuit may be achieved by using magnetic coupled reactance. When $I_1 = I_2$ then resultant flux is zero as the two coils are connected in anti-parallel. So, the inductance of the both path will be the same. If $I_1 > I_2$ then there will be a resultant flux. This flux induces the emfs in cols. 1 and 2 as shown in figure. Thus the current in path 1 is opposed and in the path 2 it is aided by the induced emfs. Therefore ,reducing the current difference in the paths.

4.3 Hybrid Model of Transistor

At low frequencies, the response of the transistor to changes of input voltage or current is instantaneous and hence the effect of shunt capacitances are neglected. But this is not in case at high frequencies.

At low frequencies, the transistors are analyzed using h-parameters. But for high frequencies, the h-parameter model is not suitable for following reasons:

- The values of h-parameters are not constant at high frequencies. Therefore, it is necessary to analyze transistor at each and every frequency which is impracticable.
- At high frequency, the h-parameters become complex in nature.

Due to the above reasons, modified T model and hybrid π models are used for high frequency analysis of the transistor.

Hybrid π Common Emitter Transistor Model

Common emitter circuit is the most important practical configuration and hence this circuit is chosen for the analysis using hybrid model shown in the figure.

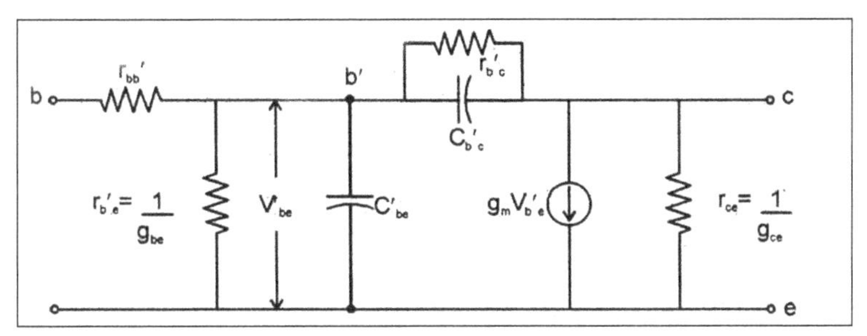

Hybrid -π Common Emitter Transistor model.

4.4 Analysis of Transistor Amplifier using h-Parameters

Hybrid parameters for a basic transistor circuit in CE configuration and its hybrid model:

- BJT is a two-port device in which one terminal is common to both the input and output ports. The behavior of the two-port network is analyzed using the current and voltage parameters at input and output ports, namely input current, input voltage, output current and output voltage. Out of these four parameters, two parameters are considered independent and the remaining two parameters are dependent. The dependent parameters are expressed in terms of the independent parameters.
- Consider the two-port network shown in the figure. The terminal behavior of any two-port network can be specified by the terminal voltages V_1 and V_2 at ports 1 and 2 respectively and currents i1 and i2entering ports 1 and 2 respectively.

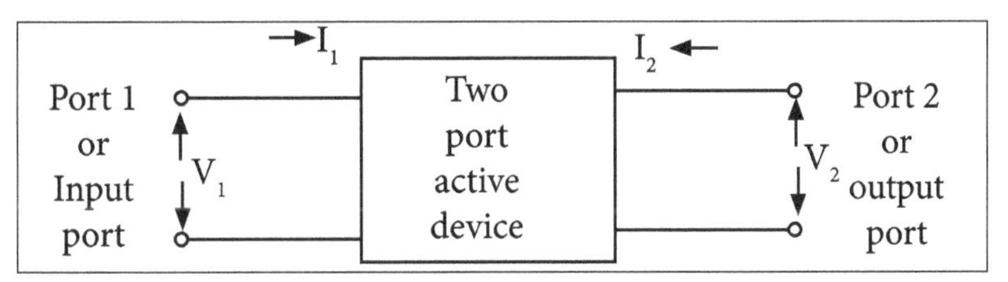

Two port network.

If the input current is i_1 and the output voltage V_2 are taken as independent variables, the input voltage V_1 and output current i_2 can be written as,

$$V_1 = h_{11} i_1 + h_{12} V_2$$

$$i_2 = h_{21} i_1 + h_{22} V_2.$$

The hybrid parameters h_{11}, h_{12}, h_{21} and h_{22} are defined as follows,

$$h_{11v} = \left[\frac{V_1}{i_1}\right] \text{ with } V_2 = 0$$

= input impedance with output port short circuited

$$h_{22} = \left[\frac{i_2}{V_2}\right] \text{ with } i_1 = 0$$

= output admittance with input port open circuited.

$$h_{12} = \left[\frac{V_1}{V_2}\right] \text{ with } i_1 = 0$$

= reverser voltage transfer ratio with input port open circuited.

$$h_{21} = \left[\frac{i_2}{i_1}\right] \text{ with } V_2 = 0$$

= forward current gain with output port short circuited.

The dimensions of h-parameters are as follows:

$$h_{11} = \Omega$$

$$h_{22} - \text{mhos } (\mho)$$

$$h_{12}, h_{21} - \text{ dimensionless}$$

As the dimensions are not alike, (i.e) they are hybrid in nature, these parameters are called as hybrid parameters.

An alternative subscript notation recommended by IEEE is commonly used:

i = 11 = input

o = 22 = output

f = 21 = forward transfer

r = 12 = reverse transfer.

In case of transistors, another subscript (b, e or c) is added to indicate the type of configuration. For example.

$h_{11e} = h_{ie}$ = Input resistance in common-emitter configuration

$h_{21b} = h_{fb}$ = Short-circuit forward current gain in common-base configuration

$h_{22c} = h_{oc}$ = Output admittance in common-collector configuration

Based on the definition of hybrid parameters, the mathematical model for two-port networks known as h-parameter model can be developed as shown in the figure:

$$V_1 = h_i i_1 + h_r V_2$$

$$i_2 = h_f i_1 + h_o V_2$$

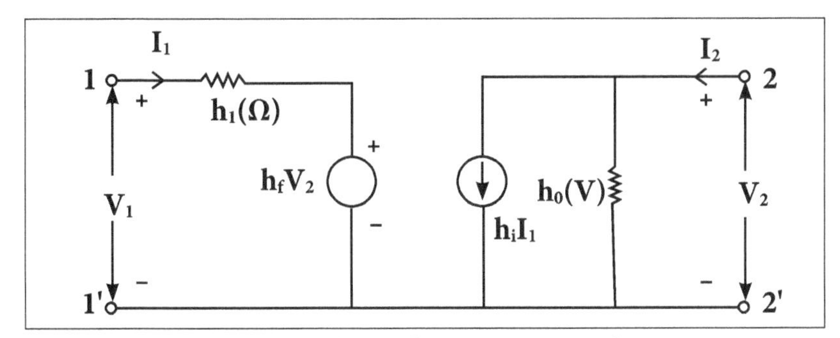

Hybrid model for the two-port network.

The model satisfies these two equations and it can be verified by writing Kirchhoff's voltage law equation in the input loop and KCL equation for the output node.

The input circuits have a dependent voltage generator and the output circuit contains a dependent current generator.

H-Parameter Model for CE Configuration

In case of common-emitter (CE) configuration, the emitter of the transistor is common to both input and output terminals. Base of the transistor is the input terminal and collector of the transistor is the output terminal. The variables are i_B, i_C, V_B (= V_{BE}) and V_C (= V_{CE}). i_B and V_C are considered as independent variables. V_B and i_C are considered as dependent variables.

$$V_B = f_1(i_B, V_C)$$

$$i_C = f_2(i_B, V_C)$$

Therefore, the equations for CE configuration can be written as,

$$V_b = h_{ie} i_b + h_{re} V_C$$

$$i_c = h_{fe} i_b + h_{oe} V_C$$

The subscript 'e' indicates that the h-parameters are for CE configuration.

h_{ie} = Input impedance when collector-emitter terminal is short circuited

h_{re} = Reverse voltage gain when base-emitter terminal is open circuited

h_{fe} = Forward current gain when the collector-emitter terminal is short circuited

h_{oe} = Output admittance when the base-emitter terminal is open circuited

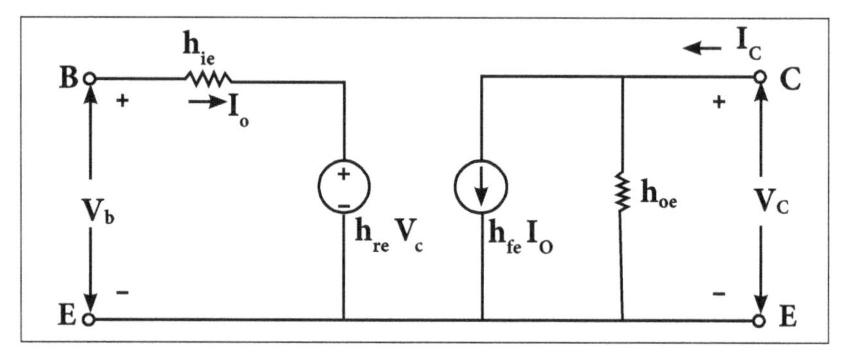

Hybrid model for transistor in CE configuration.

5

Power Semiconductor Devices

5.1 Principle of Operation and Characteristics of Thyristors

Thyristor is a three terminal device with four layers of alternating P and N type material. The three terminals are Anode, Cathode and Gate.

Thyristor is also termed as Silicon Controlled Rectifier (SCR) as it is made up of silicon and working as controlled rectifier. It is inherently a slow switching device compared to BJT's or MOSFET's because of the long carrier lifetimes used for low on-state losses and because of the large amount of stored charge. It is therefore normally used at lower switching frequencies. It has large reverse-recovery currents.

Thyristor has four unique layer construction of alternating P-type and N-type regions.

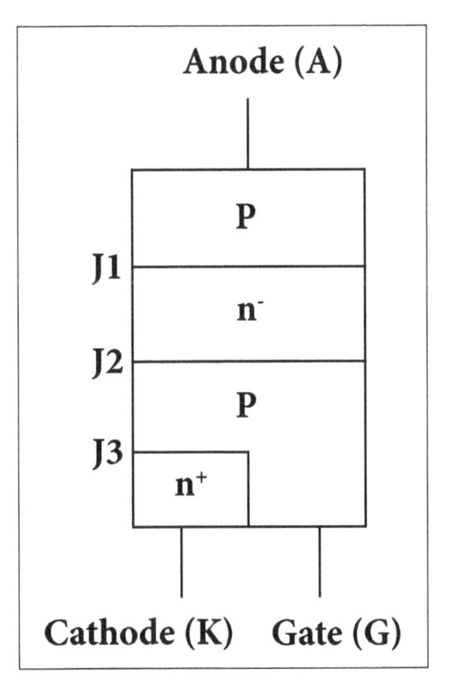

The SCR looks like two PNP transistor connected in a back to back manner. This is also called as two transistor model.

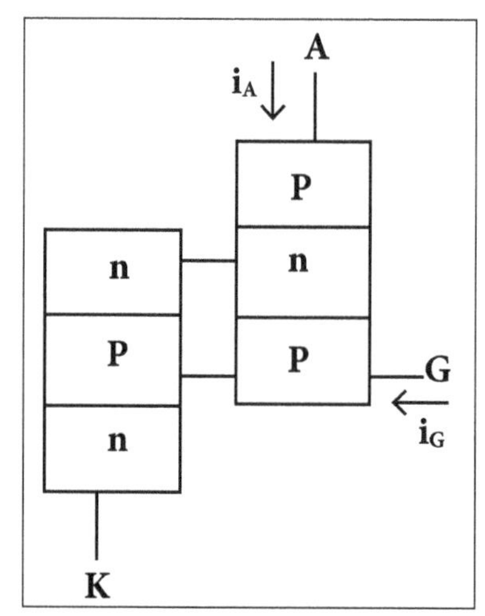

Two transistor model.

Characteristics: Static and Dynamic

Static Characteristics

The static V-I Characteristics of an SCR is shown in the figure. Thyristor V-I Characteristics are divided into three regions of operation. The three regions of operations are:

- Reverse Blocking Region.
- Forward Blocking Region.
- Forward Conduction Region.

Reverse Blocking Region

When the cathode of the SCR is made positive with respect to the anode then outer junctions J_1 and J_3 are reverse biased, whereas junction J_2 is forward biased.

Therefore small leakage current flows through the SCR and it almost remain constant with the rise in the voltage until certain voltage called avalanche breakdown voltage.

At this particular voltage, the depletion regions of the junctions J_1 and J_3 are broken down and the thyristor conducts in the reverse bias direction. At certain voltage, current flowing through the thyristor increases suddenly to high value.

Forward Blocking Region

In this region anode is made positive with respect to the cathode and therefore junction J_1 and J_3 are the forward biased and the junction J_2 is reversed biased.

Hence the thyristor is in forward blocking condition and small leakage current flows during this condition due to the drift of charge carriers. In this condition device does not conduct.

Forward Conduction Region

When the thyristor is in forward blocking condition and anode to cathode voltage is increased further and gate circuit is in open condition then at particular voltage called as forward break over voltage (V_{bo}), avalanche breakdown occurs and thyristor starts conducting in forward mode.

The voltage starts falling from some hundreds of volts to 1-2 volts and suddenly very large current starts flowing through the SCR. The process of conducting the thyristor damages the device.

In order to conduct the thyristor safely, gate current I_g is provided that results in the reduction in the forward break over voltage less than (V_{bo}). With increase in the gate current I_g the conduction voltage is reduced.

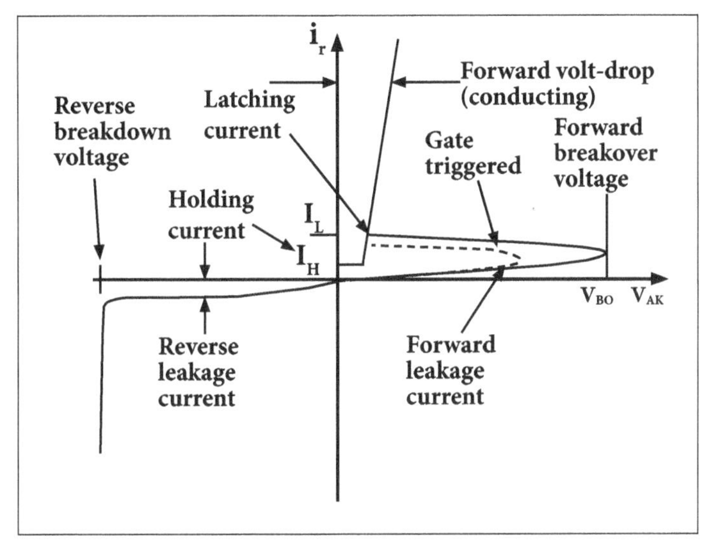

Static V-I Characteristics.

Dynamic Characteristics

The time variation of voltage across the thyristor and current through it during turn ON and turn OFF process gives the dynamic or switching characteristic of SCR.

Switching Characteristic During Turn ON Time

Turn ON Time

It is the time during which it changes from forward blocking state to ON state. Total turn ON time is divided into 3 intervals:

- Delay time.
- Rise time.

- Spread time.

Delay Time

If I_g and I_a represent the final value of gate current and anode current then the delay time can be explained as time during which the gate current attains $0.9I_g$ to the instant anode current reaches $0.1I_g$ or the anode current rises from forward leakage current to $0.1I_a$.

Gate current $0.9I_g$ to $0.1I_a$.

Anode voltage falls from V_a to $0.9V_a$. Anode current rises from forward leakage current to $0.1\ I_a$.

Rise Time (tr)

Time during which the anode current rises from $0.1I_a$ to $0.9I_a$. Forward blocking voltage falls from $0.9V_a$ to $0.1V_a$. V_a is the initial forward blocking voltage.

Spread Time (tp)

Time taken by the anode current to rise from $0.9I_a$ to I_a. Time for the forward voltage to fall from 0.1V to ON state voltage drop of 1 to 1.5V.

During turn ON, SCR is considered to be a charge controlled device. A certain amount of charge is injected in the gate region to begin conduction. So higher the magnitude of gate current, it requires less time to inject the charges. Thus turn ON time is reduced by using large magnitude of gate current.

Switching Characteristics During Turn Off

Thyristor turn OFF means it changed from ON to OFF state. Once thyristor is ON there is no role of gate. Thyristor can be made to turn OFF by reducing the anode current below the latching current.

Here we assume the latching current to be zero. If a forward voltage is applied across the SCR at the moment it reaches zero then SCR will not be able to block this forward voltage. Because the charges trapped in the 4-layer are still favorable for conduction and it may turn ON the device. So to avoid such a case, SCR is reverse biased for some time even if the anode current has reached to zero.

So now the turn OFF time can be different as the anode current becomes zero to the instant when SCR regains its forward blocking capability.

$$t_q = t_{rr} + t_{qr}$$

Where,

t_q is the turn OFF time, t_{rr} is the reverse recovery time, t_{qr} is the gate recovery time

At t_L anode current is zero. Now anode current builds up in reverse direction with same dv/dt slope. This is due to the presence of charge carriers in the four layers. The reverse recovery current removes the excess carriers from J_1 and J_3 between the instants t_1 and t_3.

At instant t_3 the end junction J_1 and J_3 is recovered. But J_2 still has trapped charges which decay due to recombination so the reverse voltage has to be maintained for some more time. The time taken for the recombination of charges between t_3 and t_4 is called gate recovery time t_{qr}. Junction J_2 recovered and now a forward voltage can be applied across SCR.

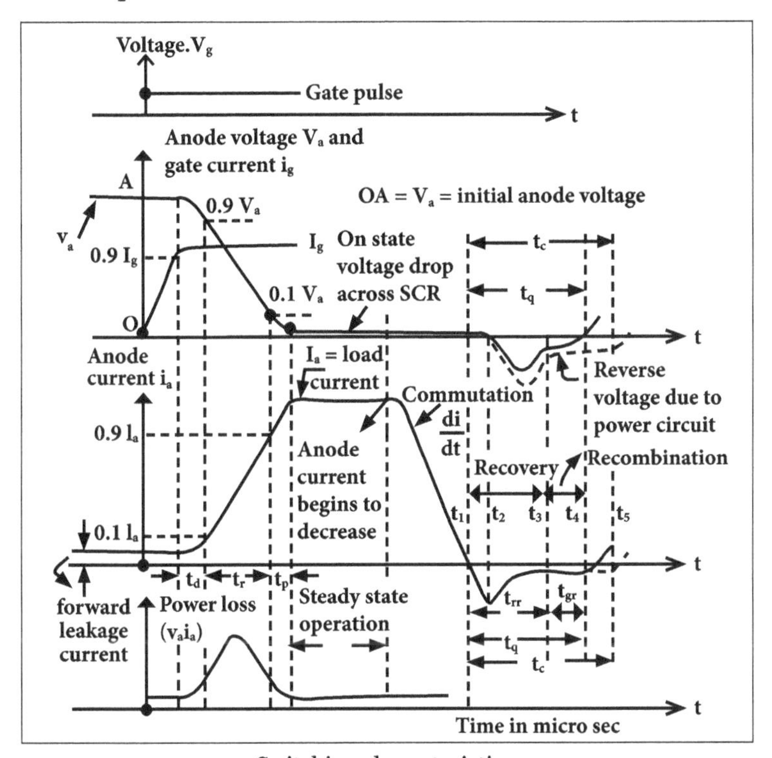

Switching characteristics.

Thyristor Protection

For reliable operation of SCR, it must be operated within the specific ratings.

SCRs are very delicate devices and so they has to be protected against abnormal operating conditions. Various protection of SCR are:

- dv/dt protection.
- di/dt protection.

- Over voltage protection.
- Over current protection.

5.1.1 Silicon Control Rectifiers

SCR Equivalent Circuit

A Silicon-Controlled Rectifier (SCR) is a four-layer (p-n-p-n) semiconductor device that doesn't allow current to flow until it is triggered and once triggered, it will only allow the flow of current in one direction.

It has three terminals:

- An input control terminal referred to as a 'gate'.
- An output terminal known as the 'anode'.
- A terminal known as a 'cathode', which is common to both the gate and the anode.

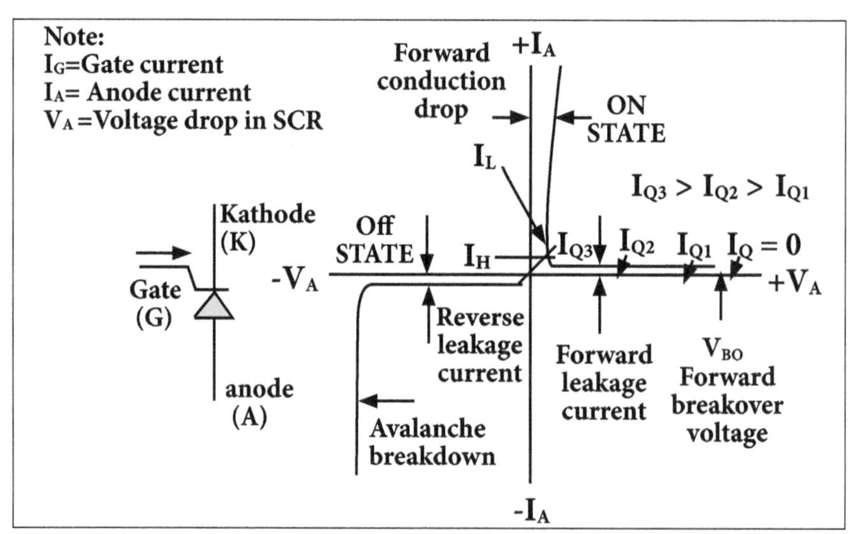

SCR characteristics.

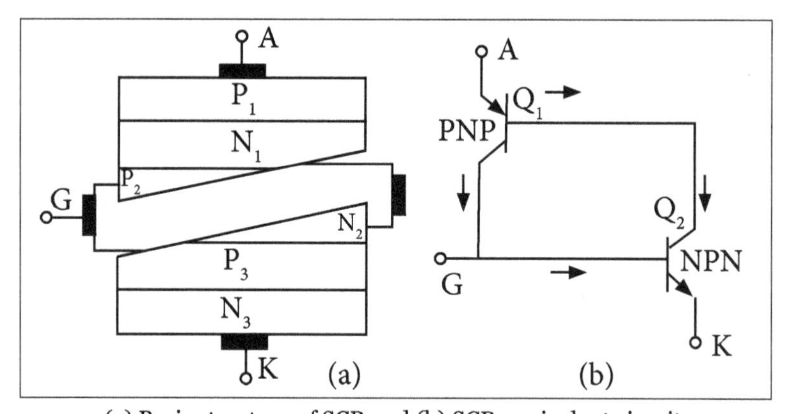

(a) Basic structure of SCR and (b) SCR equivalent circuit.

The basic structure of SCR can be divided into 3-layer structure as shown in figure (a). It may be noted that the upper 3-layer structure of a PNP transistor, whereas the lower one is an NPN transistor. Thus, SCR can be represented by two transistors Q_1 (i.e., PNP) and Q_2 (i.e., NPN) interconnected as shown in figure (b). Sometimes, the circuit is also called as the two transistor analogs or an ideal latch of a SCR.

It is evident from the figure (b) that the collector of each transistor is connected to the base of other transistor. Therefore, the collector current of transistor Q_1 is the base current of transistor Q_2 and the base current of transistor Q_1 is the collector current of transistor Q_2.

For this circuit, we have an action of positive feedback or regeneration. It means that if there is a change in current, at any point in the loop, it is amplified and returned to the starting point with the same phase.

For example, if the base current of transistor Q_3 will also increase. It causes more base current through transistors Q_1 due to which the collector current of transistor Q_1 increases. This action will continue till both the transistors are driven into saturation. In this case, the SCR acts like a ON switch and it will pass the current from anode to cathode.

On the other hand, if the base current of transistor Q_2 decreases, the collector current of transistor Q_2 will also decrease. It causes the reduced based current through transistor Q_1 due to which the collector current of transistor Q_1 decreases. This action will continue till both the transistors are driven into act-off. In this case, the SCR acts like a OFF switch and hence it will block the current from anode to cathode.

Holding Current

This is the value of anode current below which the SCR switches from ON state to OFF state. The value increases with decreasing values of I_G and is maximum for IG =0 and also by increasing the external circuit resistance.

5.2 Power IGBT and Power MOSFET

Power IGBT

Insulated Gate Bipolar Transistor also called as IGBT is a cross between a conventional Bipolar Junction Transistor and a Field Effect Transistor (MOSFET) making it ideal as a semiconductor switching device.

This IGBT transistor takes the best parts of these two types of transistors, the high input impedance and switching speed of the MOSFET with the low saturation voltage of a bipolar transistor, and combines them together to produce the another type of the transistor switching device that is capable of handling large collector-emitter currents with zero gate current drive.

The Insulated Gate Bipolar Transistor (IGBT) uses the insulated gate technology of the MOSFET with the output performance characteristics of the conventional bipolar transistor. The result of this hybrid combination is that the "IGBT Transistor" has the output switching and conduction characteristics of a bipolar transistor but is voltage-controlled like a MOSFET.

It is mainly used in power electronics applications such as inverters, converters and power supplies. High-current and high-voltage bipolar are available, but their switching speeds are slow, while power MOSFETs may have high switching speeds, but high current and high voltage devices are expensive and hard to achieve.

The advantage gained by the insulated gate bipolar transistor device over a BJT or MOSFET is that it offers great power gain than the bipolar type together with its higher voltage operation and lower input loss of the MOSFET.

Insulated Gate Bipolar Transistor

IGBT

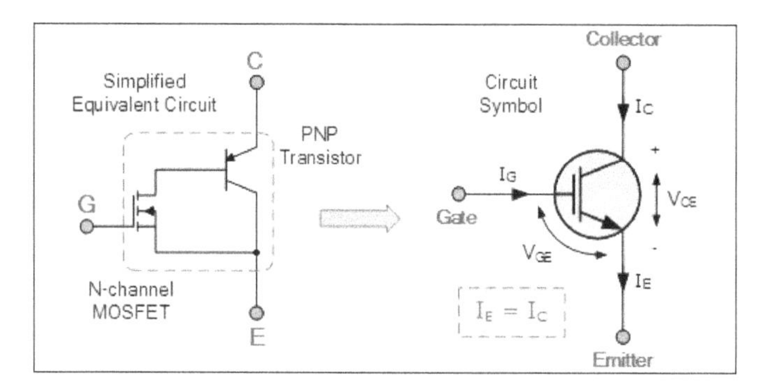

The insulated gate bipolar transistor is a three terminal, trans conductance device which combines an insulated gate N-channel MOSFET input with the PNP bipolar transistor output connected in the type of Darlington configuration. The terminals are labelled as:

- Collector,
- Emitter,
- Gate.

Two of its terminals (C-E) are associated with a conductance path and its third terminal (G) associated with its control. This amount of amplification can be achieved by the insulated gate bipolar transistor is a ratio between its output signal and its input signal.

For a conventional bipolar junction transistor (BJT), the amount of gain is nearly equal to the ratio of the output current to the input current, known as Beta. For a metal oxide semiconductor field effect transistor or MOSFET, it has no input current as the gate is isolated from the main current carrying channel.

Therefore, the FET's gain is equal to the ratio of the output current change to the input voltage change, making it a trans conductance device. Then we can treat the IGBT as a power BJT for which the base current is provided by a MOSFET.

The Insulated Gate Bipolar Transistor can be used in small signal amplifier circuits in the same way as the BJT or MOSFET type transistors. As the IGBT combines low conduction loss of a BJT with the high switching speed of the power MOSFET, an optimal solid state switch exists which is ideal for use.

An insulated gate bipolar transistor is generally turned "ON" or "OFF" by activating and deactivating its Gate terminal. The constant positive voltage input signal across the Gate and the Emitter will be in "ON" state, while removal of the input signal causes it to turn "OFF" in the same way as a bipolar transistor or MOSFET.

IGBT Characteristics

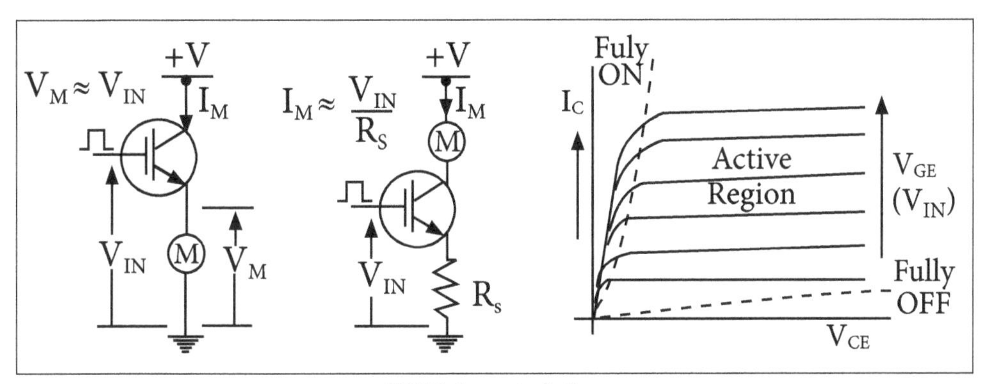

IGBT characteristics.

IGBT is a voltage-controlled device, it requires only a small voltage on the Gate to maintain conduction through the device unlike BJT's which require that the Base current is supplied continuously in the sufficient quantity to maintain the saturation.

Also the IGBT is a unidirectional device, which means it can switch current only in the "forward direction", that is from Collector to Emitter. But, MOSFET have bi-directional current switching capabilities which can be controlled in the forward direction and uncontrolled in the reverse direction.

The principal of operation and Gate drive circuits for IGBT are very similar to the N-channel power MOSFET. The basic difference is that the resistance offered by the main conducting channel when current flows through this device in its "ON" state is very much smaller in the IGBT. Because of this, the current rating is much higher when compared with an equivalent power MOSFET.

The main advantages of using the Insulated Gate Bipolar Transistor over other types of transistor devices are its high voltage capability, ease of drive, low ON-resistance, relatively fast switching speeds and zero gate drive current.

It is a good choice for moderate speed, high voltage applications such as in pulse-width modulated (PWM), variable speed control, switch-mode power supplies or solar powered DC to AC inverter and frequency converter applications operating in the hundreds of kilohertz range. The general comparison between BJT's, MOSFET's and IGBT's is given in the following table.

Power MOSFET

A Metal-Oxide-Semiconductor Field-Effect Transistor (MOSFET) is one that is developed by combining the areas of field-effect concept and MOS technology. Conventional planar MOSFET has the restriction of handling high power. In high power applications, the VMOS or Double-diffused vertical MOSFET is used which is simply known as Power MOSFET.

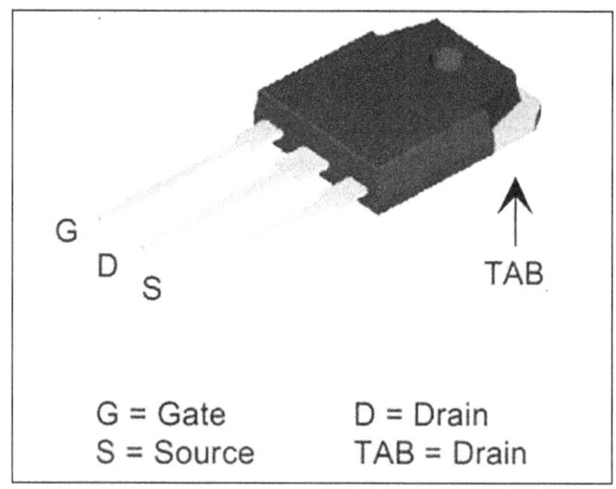

Power MOSFET.

The Power MOSFET is the three terminal four layer, unipolar semiconductor device.

The MOSFET is a majority carrier device and as the majority carriers have no recombination delays, MOSFET achieves very high bandwidths and switching times.

The gate is electrically isolated from source and while this provides the MOSFET with high input impedance even forms a good capacitor.

MOSFETs does not have secondary breakdown area, their drain to source resistance has a positive temperature coefficient so they tends to be self-protective. It has very low ON resistance and has no junction voltage drop when forward biased. This feature makes MOSFET an extremely attractive power supply switching device.

Symbol

The symbol for n-channel MOSFET is given below. The direction of the arrow on the head that goes to the body region indicates the direction of current flow. Since this is the symbol for n channel MOSFET, the arrow is inwards. For p-channel MOSFET, the arrow will be towards outside.

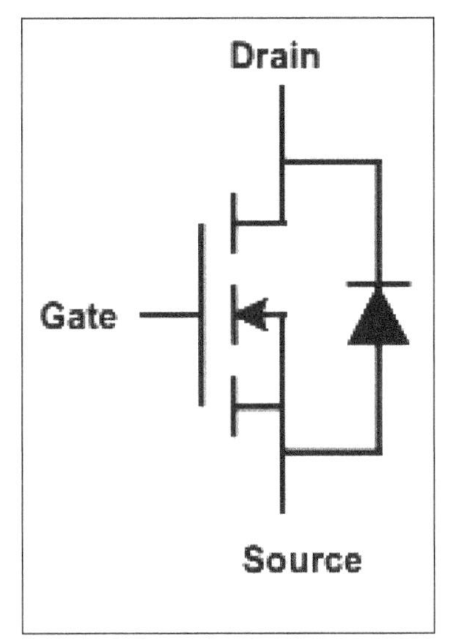

Symbol for n-channel MOSFET.

Structure

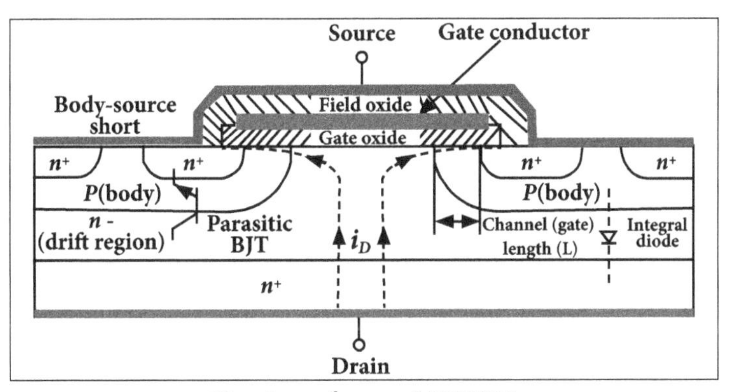

Structure of power MOSFET.

The Power MOSFET has a vertically oriented four layer structure of alternating P and N type (n^+ pn n^+) layers.

The P type middle layer is called as body of MOSFET. In this region, channel is formed between source and drain. The n- layer is called as drift region which determines the breakdown voltage of the device. This n- region is present only in Power MOSFETs not in signal level MOSFET. The gate terminal is isolated from body by silicon dioxide layer.

When the positive gate voltage is applied with respect to source, n-channel is formed between the sources to drain.

As shown in the figure, there is a parasitic npn BJT between source and drain. To avoid this BJT turns ON, the p-type body region is shorted to source region by overlapping the source metallization on to the p type body. The result is a parasitic diode which is formed between drain to source terminals. This integral diode plays an important role in half and full bridge converter circuits.

Characteristics

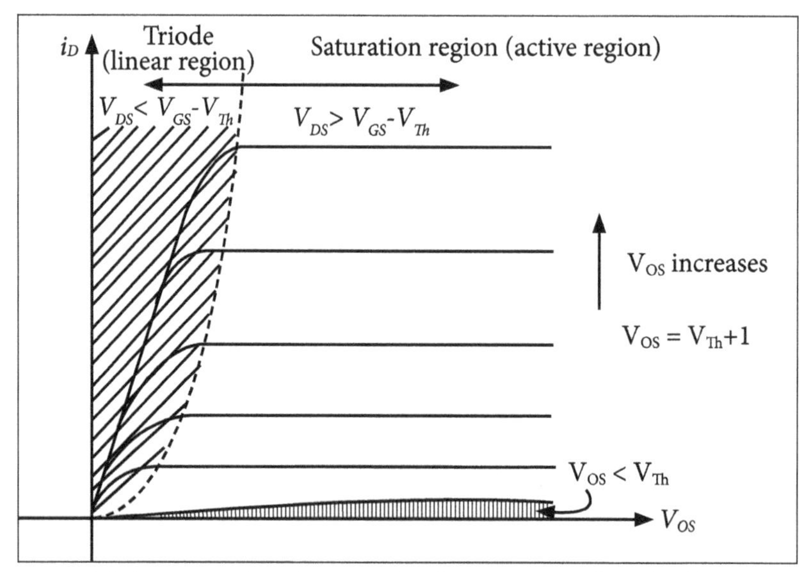

VI characteristics of n-channel enhancement mode MOSFET.

Comparison of Power Devices

Device Characteristic	Power Bipolar	Power MOSFET	Power IGBT
Voltage Rating	High < 1 KV	High < 1 KV	Very High > 1kV
Current Rating	High <500 A	Low < 200 A	High > 500 A
Input Drive	Current 20-200 h_{FE}	Voltage V_{GS} 3-10 V	Voltage V_{GE} 4-8 V
Input Impedance	Low	High	High
Output Impedance	Low	Medium	Low
Switching Speed	Slow (uS)	Fast (n S)	Medium
Cost	Low	Medium	High

5.3 FET

The Field Effect Transistor or FET uses the voltage that is applied to their input terminal, called as Gate, to control the current flowing through them resulting in the output current being proportional to the input voltage. Their operation relies on an electric field generated by the input gate voltage hence it is named as field effect.

It is an unipolar device i. e. operation depends on only one type of the charge carriers (h or e). It is a Voltage Controlled Device, hence gate voltage controls the drain current.

5.3.1 JFET Characteristics (Qualitative Explanation)

Characteristics

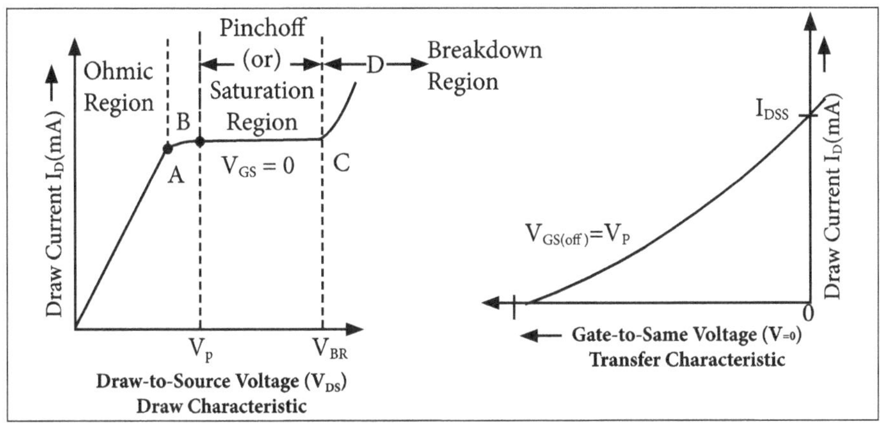

Drain characteristics Transfer characteristics.

Characteristics of p-Channel JFET

There are two important characteristics of a JFET;

- Transfer characteristics.
- Drain characteristics.

Transfer Characteristics

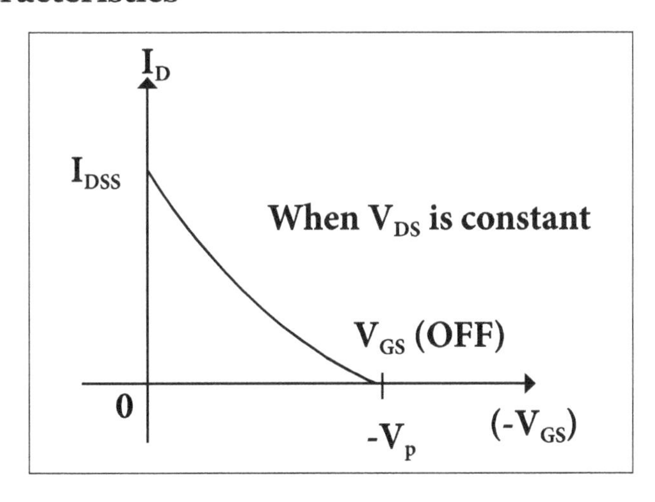

The upper end of the curve as shown by the drain current value is equal to IDSS. (i.e) When $V_{GS} = 0$, the drain current is maximum while the lower end is indicated by a voltage equal to $V_{GS(off)}$ or V_p. The curve is a part of the parabola. Hence, it may be expressed as,

$$I_D = I_{DSS}\left(1 - \frac{V_{GS}}{V_P}\right)^2$$

Drain Characteristics

When $V_{DS} = 0$, there is no attracting potential at the drain and hence drain current $I_D = 0$, even the channel between the gates is fully open. As V_{DS} is increased, the drain current I_D increases linearly. This region is called ohmic region.

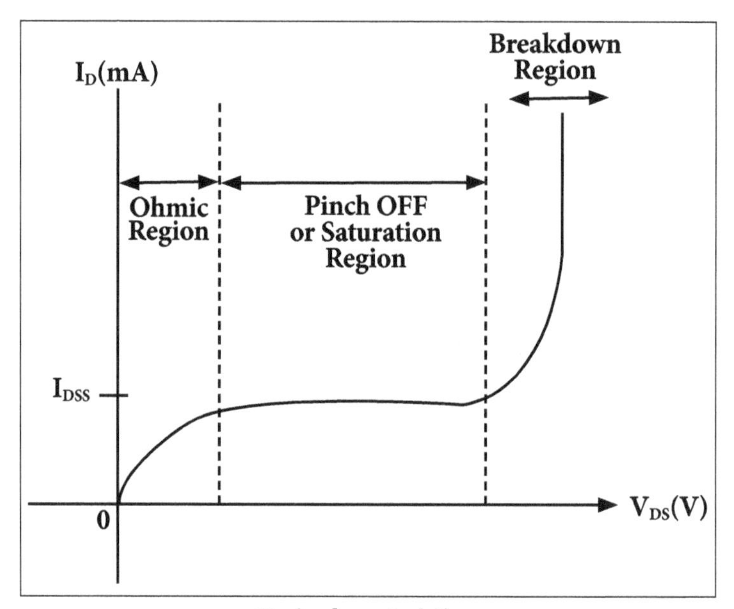

Drain characteristics.

In the saturation region, drain current remains constant at its maximum value IDSS. The drain current in the pinch off region depends upon the gate to source voltage and is given by the relation,

$$I_D = I_{DSS}\left(1 - \frac{V_{GS}}{V_P}\right)^2$$

This relation is known as "Schottky Equation".

In the breakdown region, the drain current increases rapidly as the drain to source voltage is increased. It is because of the breakdown of gate to source junction due to avalanche effect.

5.3.2 MOSFET Characteristics

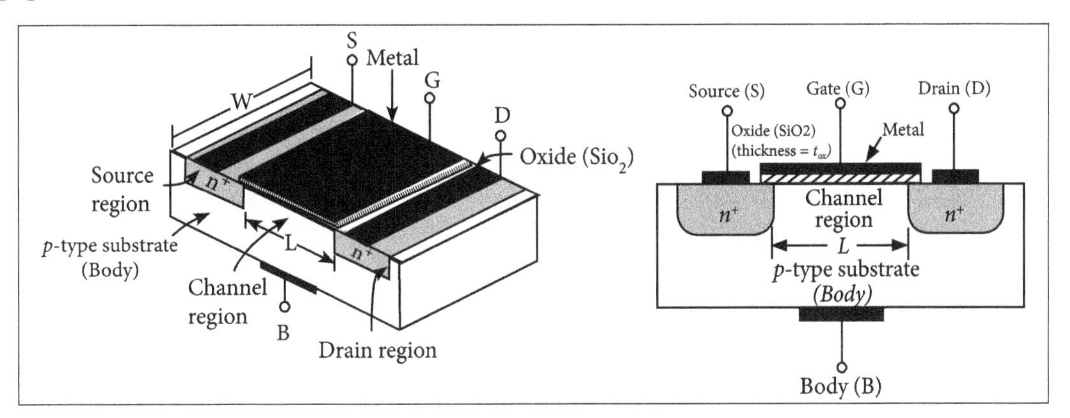

"MOS" represents metal-oxide-semiconductor structure. MOSFET is a four terminal device. The four terminals of the device are:

- Gate (G).
- Source (S).
- Drain (D).
- Body (B).

The device size is mentioned by channel width (W) and channel length (L). There are two kinds of MOSFETs which are n-channel and p-channel devices. The device structure is basically symmetric in terms of drain and source terminals. Source and drain terminals are specified by the operation voltage.

Operation with Zero Gate Voltage

The MOS structure forms a parallel- plate capacitor with gate oxide layer in the middle. The two pn junctions (S - B and D - B) are then connected as back to back diodes. The source and the drain terminals are isolated by two depletion regions that do not conduct current. The operating principles may be introduced by using the n-channel MOSFET as an example.

Creating a Channel for Current Flow

Positive charges accumulated in gate as a positive voltage applies to the gate electrode. Electric field is forming a depletion region by means of pushing holes in p-type substrate away from the surface. The electrons accumulated on the substrate surface as gate voltage exceeds a threshold voltage V_t. The induced n region thus forms a channel for the current flow from drain to the source. The substrate is inverted from p -type to n - type inversion layer, in order to create the channel. The field controls the amount of charge in the channel and it helps in determining the conductivity of the channel.

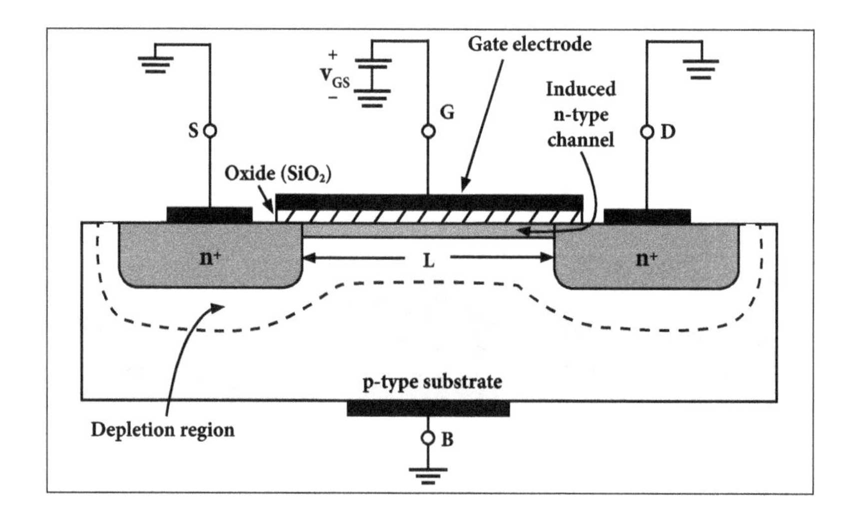

Applying a Small Drain Voltage

A positive $v_{GS} > V_t$ is used to induce the channel → n - channel enhancement -type MOSFET. Free electrons are travelled from the source to drain through the induced n-channel due to a small v_{DS}. The current i_D flows from drain to source. Which is proportional to number of carriers in the induced channel. The channel is being controlled by the effective voltage or the overdrive voltage: $v_{OV} = v_{GS} - v_t$ The electron charge in the channel due to the overdrive voltage is given by,.

Gate oxide capacitance C_{ox} is referred to as capacitance per unit area MOSFET and it can be approximated as a

Linear resistor in this region with a resistance value is inversely proportional to the excess gate voltage.

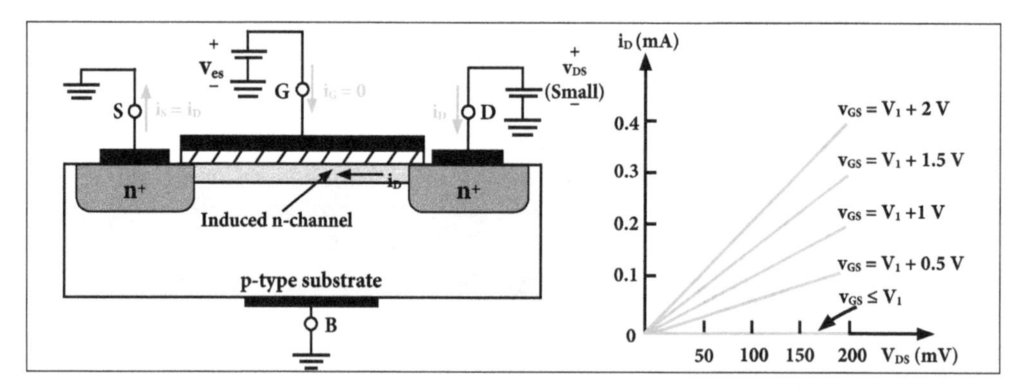

Operation as Increasing Drain Voltage

As v_{DS} increases, the voltage in the channel increases from 0 to v_{DS}. The voltage between the gate and the points along the channel decreases from v_{GS} at the source end to $(v_{GS}\ v_{DS})$ at the drain end. As the inversion layer depends on the voltage difference across the MOS structure, increase in the v_{DS} will result in a tapered channel. The resistance increases due to tapered channel and the $i_D - v_{DS}$ curve is not a straight line from now. At the point $v_{DSsat} = v_{GS} - V_t$ the channel is pinched off at the drain side. If v_{DS} is increased beyond this value, will result in causing slight effect on the channel shape and i_D saturates at this value. Triode region: $v_{DS} < v_{DSsat}$.

Saturation region: $v_{DS} \geq v_{DSsa}$.

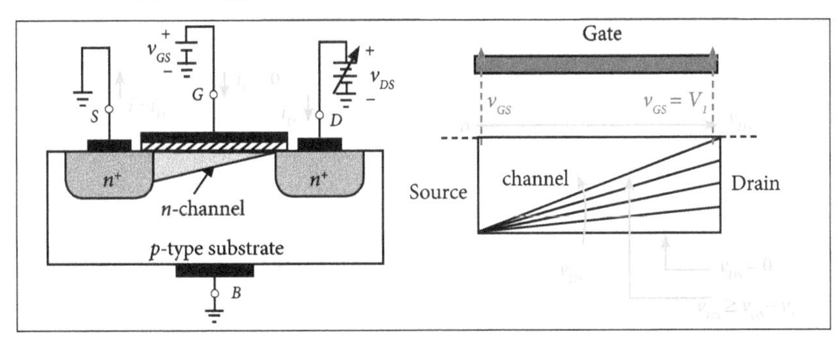

V-I Characteristics

The MOSFET is a three terminal device. The voltage on the gate terminal controls the flow of current between the output terminals, Source and the Drain. The source terminal is common between the input and the output of MOSFET. The output characteristic of a MOSFET is a plot of drain current (i_D) as a function of the Drain –Source voltage (v_{DS}) against the gate source voltage (v_{GS}) as a parameter. The figure explains the I-V characteristics of MOSFET.

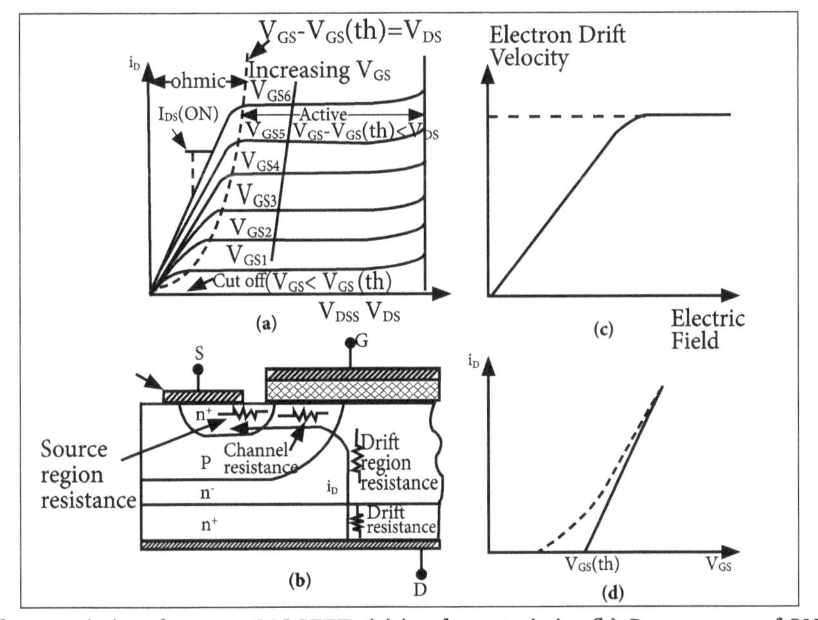

Output i-v characteristics of a Power MOSFET; (a) i-v characteristics;(b) Components of ON-state resistance; (c) Electron drift velocity vs Electric field; (d) Transfer.

The current through the channel is I = V/R, Where V is the DRAIN – SOURCE voltage. We are assuming that V << VT here. The channel resistance, R is given by,

$$R = \frac{L}{qn\mu aW} = \frac{L}{qn_s\mu W}$$

Where,

$$n_s = (c_i / q) \times (V_{GS} - V_T)$$

The channel current is,

$$I = V(qn_s\mu W)/L = Vq\mu W(c_i/q) \times (V_{GS} - V_T)/L$$

$$I = \mu W c_i \times (V_{GS} - V_T)V/L$$

5.3.3 Static and Transfer (Enhancement and Depletion Mode)

The MOSFET is used to represent Metal-Oxide Semiconductor Field Effect Transistor. Like JFET, it has a source, gate and a drain. However, unlike JFET, the gate of a MOSFET is insulated from the channel. Because of this, the MOSFET is sometimes called as an IGFET which stands for Insulated Gated Field Effect Transistor.

D-MOSFET

The depletion types MOSFET are operated in two different modes as given below:

- Depletion mode: The device operates in this mode, when and if the gate voltage is negative.
- Enhancement mode: The device operates in this mode, when and if the gate voltage is positive.

As the depletion type MOSFET can be operated in depletion enhancement mode, this device is commonly termed as depletion enhancement (DE) type MOSFET.

The working of the entire structure of the device as a parallel plate capacitor has one of the plate is formed by the gate and the other by the semiconductor channel. The plates are separated by a dielectric (Sio2layer). We know that if one plate of a capacitor is made negative, it induces a positive charge on the opposite plate and vice-versa.

Depletion Mode

The figure shows a MOSFET with a negative gate to source voltage. The negative voltage on the gate induces a positive charge in the channel. As this is the case, free electrons in the vicinity of positive charge are replied away in the channel. As a result of this, the channel is depleted of free electrons. This reduced the number of free electrons passing through the channel. Hence, as the value of negative gate to source voltage is increased the value of drain voltage known as V_{GS} (off), the channel totally depletes all the free electrons and thus the drain current reduces to zero. Therefore with the negative gate voltage, the operation of MOSFET is similar to that of a JFET.

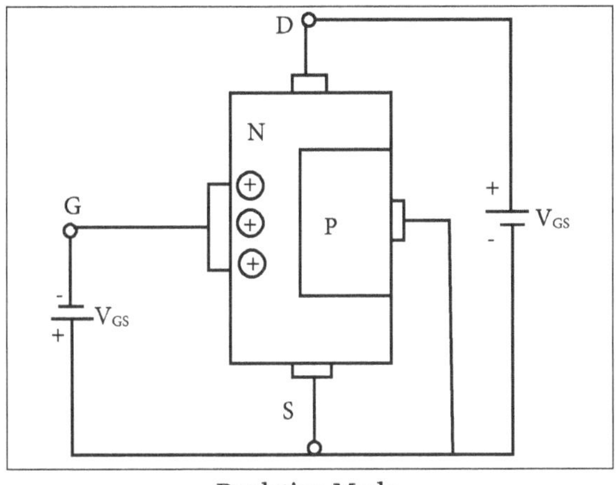

Depletion Mode.

It is evident that negative gate voltage depletes the channel of free electrons. It is because of the fact that the working of a MOSFET with a negative gate voltage is known as depletion mode.

Characteristics

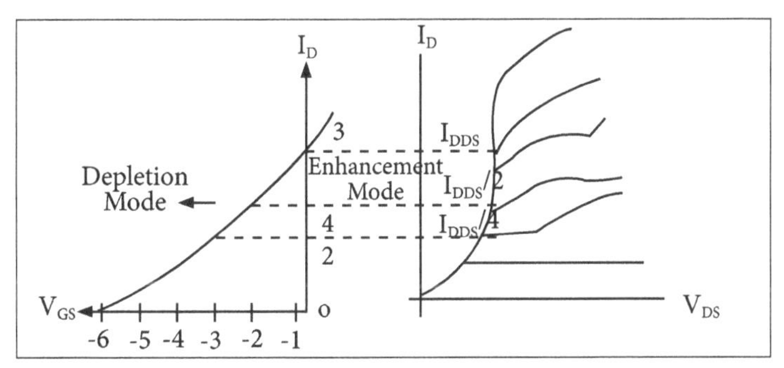

Transfer and Drain Characteristic of Depletion n-Channel MOSFET.

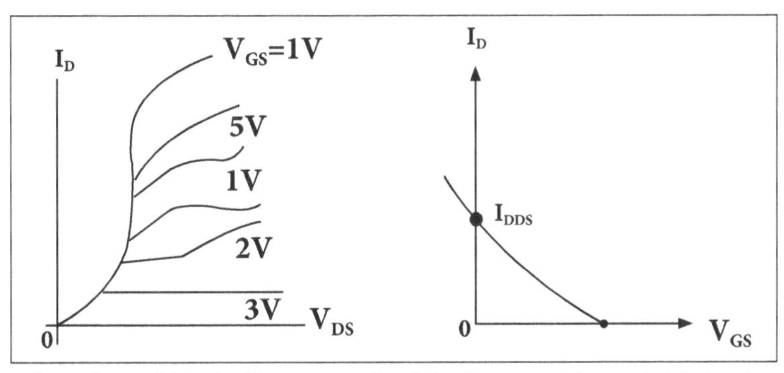

Transfer and Drain Characteristic of Depletion p-Channel MOSFET.

E-MOSFET

Construction of an E-MOSFET

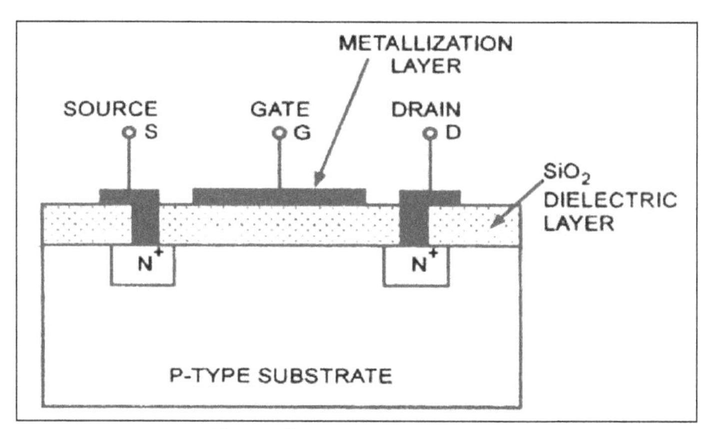

N-Channel E-MOSFET Structure.

Construction of E-MOSFET

Figure shows the construction of an N-channel E-MOSFET. The important difference between the construction of DE-MOSFET to that of E-MOSFET, E-MOSFET substrate extends all the way to the silicon dioxide (SiO_2) and channels are not doped between the source and the drain. Channels are electrically induced in these MOSFETs, when a positive gate-source voltage V_{GS} is applied to it.

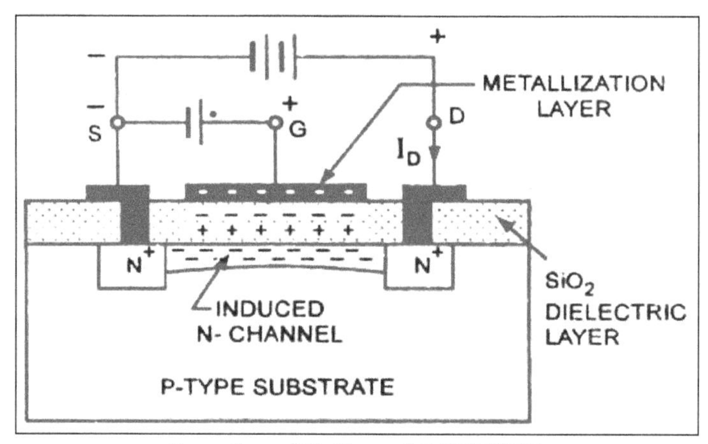

Operation of N-Channel E-MOSFET.

Working of an E-MOSFET

This MOSFET operates only in the enhancement mode and has no depletion mode. It operates with large positive gate voltage only. When the gate-source voltage $V_{GS} = 0$, it does not conduct. This is the reason that it is called as normally-off MOSFET. In these MOSFET's drain current I_D flows only when V_{GS} exceeds V_{GST}.

When drain is being applied with positive voltage with respect to the source and no potential is applied to the gate, two N-regions and one P-substrate from two P-N junctions are connected back to back with the resistance of the P-substrate. Hence, a very small drain current flows. If the P-type substrate is connected to the source terminal, there is zero voltage across source substrate junction and drain substrate junction remains reverse biased.

When the gate is changed to positive with respect to source and the substrate, negative charge carriers within the substrate are being attracted to the positive gate and accumulate close to the surface of the substrate. On increasing the gate voltage, more and more electrons are accumulated under the gate. Since these electrons cannot flow across the insulated layer of the silicon-di-oxide to the gate, they can accumulated at the surface of the substrate immediately below the gate. These accumulated minority charge carriers causes the N -type channel to stretch from drain to the source. When this happens, a channel is induced by forming what is termed as an inversion layer (N-type). a drain current starts flowing. The strength of the drain current depends upon channel resistance which, again, depends upon the number of the charge carriers attracted to positive gate. Hence, the drain current is now controlled by the gate potential.

As the conductivity of the channel is enhanced by the positive bias on the gate, this device is also known as the enhancement MOSFET or E- MOSFET.

Gate-to-source threshold voltage V_{GST} is a voltage which is the minimum value of gate-to-source voltage V_{GS} that is needed to form the inversion layer (N-type). For the value of V_{GS} below V_{GST}, the drain current $I_D = 0$. Whereas for V_{GS} exceeding V_{GST}, an N-type inversion layer connects the source to the drain and the drain current I_D is large. V_{GST} may vary from less than 1 V to more than 5 V, depending upon the device being used.

JFET's and DE-MOSFET's are classified as the depletion-mode devices as their conductivity depends on the action of the depletion layers. E-MOSFET is classified as the enhancement-mode device as its conductivity depends on the action of the inversion layer. Depletion mode devices are usually ON when the gate-source voltage $V_{GS} = 0$, whereas the enhancement-mode devices are usually OFF when $V_{GS} = 0$.

Characteristics of an E-MOSFET

Drain Characteristics of E-MOSFET

Drain characteristics of an N-channel E-MOSFET are shown. The lowest curve is the V_{GST} curve. I_D is approximately zero. When V_{GS} is lesser than V_{GST}. When V_{GS} is greater than V_{GST}, the device turns-on and the drain current I_D is controlled by the gate voltage.

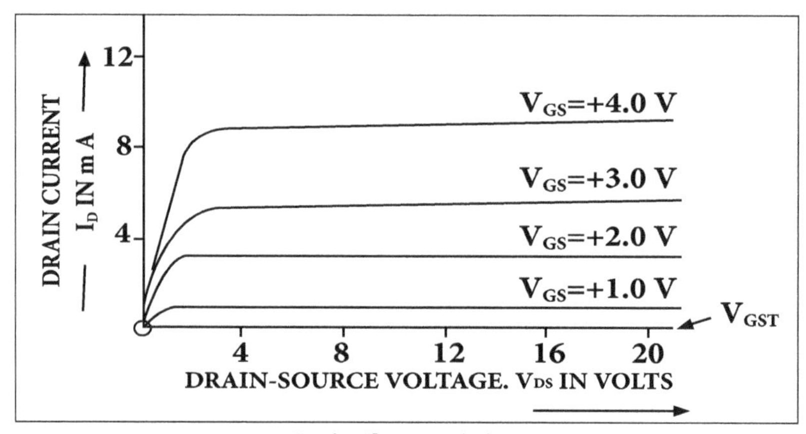

Drain Characteristics.

The characteristic curves have almost vertical and horizontal parts. The almost vertical components of the curves corresponds to the ohmic region and the horizontal components corresponds to the constant current region. Hence, E-MOSFET can be operated in either of the regions i.e. it can be used as the variable-voltage resistor or as the constant current source.

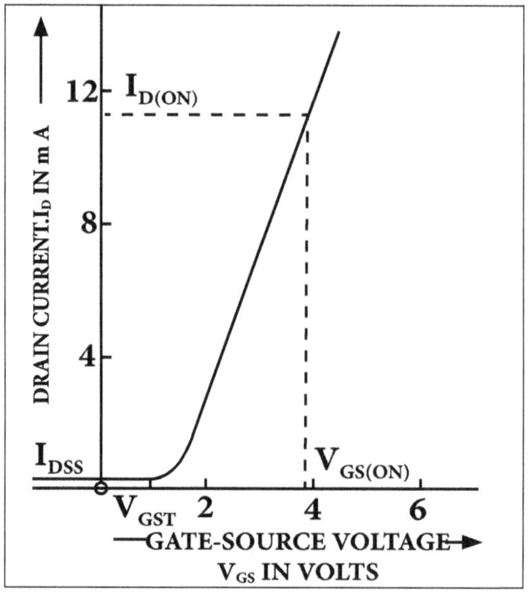

Transfer Characteristic.

E-MOSFET's Transfer Characteristics

The above figure shows the typical Trans conductance curve. The current IDSS at V_{GS} <=0 is very small, of the order of a few Nano-amperes. On making the V_{GS} positive, the drain current I_D increases slowly initially and then with a rapid increase in V_{GS}. The manufacturer sometimes indicates the gate-source threshold voltage V_{GST} at which the drain current I_D attains some defined small value, let's say 10μA.

A current $I_{D(ON)}$, corresponding approximately to the maximum value given on the drain characteristics and the values of V_{GS} required to give this current $V_{GS}(ON)$ are also usually given. However it does follow a similar "square law type" of relationship. The equation for the transfer characteristics of E-MOSFET's is given as,

$$I_D = K(V_{GS} - V_{GST})2$$

Schematic Symbols of E-MOSFET

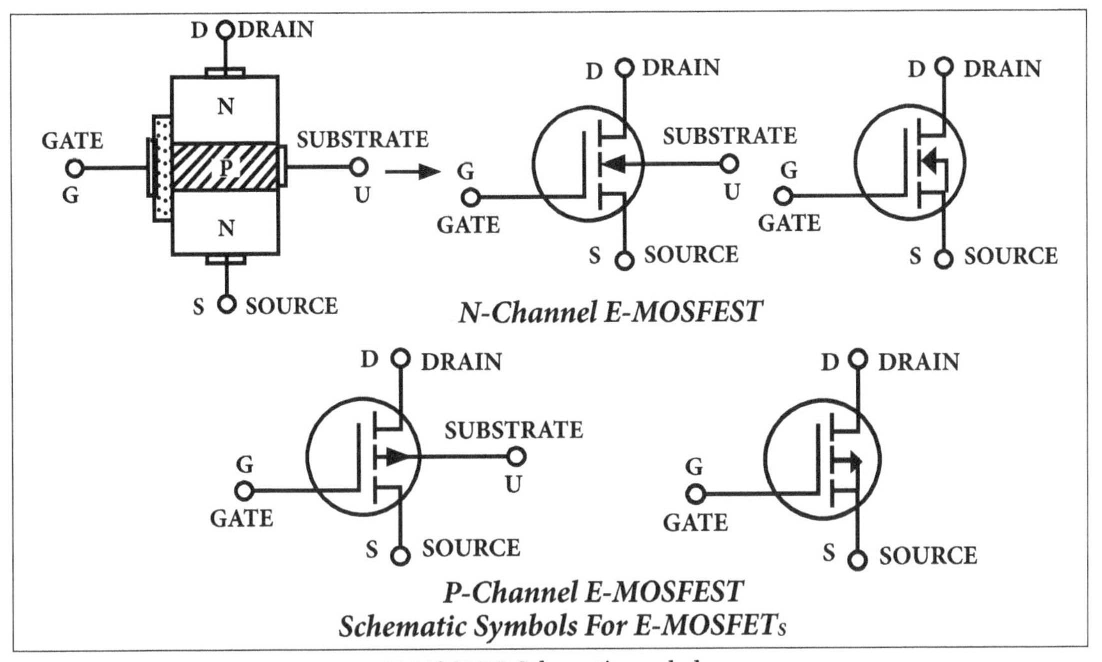

E-MOSFET-Schematic symbols.

Schematic symbols for an N-channel E-MOSFET are shown in the figure. For zero value of V_{GS}, the E-MOSFET is OFF because there is no conducting channel between source and drain. Each of the schematic symbols shown in figures has broken channel line to indicate this normally OFF condition. As we know that for V_{GS} exceeding the threshold voltage V_{GST}, an N-type inversion layer connecting the source to drain is created.

In each of the schematic symbols, the arrow points to this inversion layer, which acts like an N-channel when the device is conducting. In each case, the fact that the device has an insulated gate is indicated by the gate not making direct contact with the channel. The schematic symbol shown in the above figure shows the source and substrate internally connected, while the other symbol shown in the figure shows the substrate connection brought out separately from the source.

The schematic symbols for a P-channel E-MOSFET are also shown in the figure. In these cases, the arrow points outwards.

5.3.4 Low Frequency Model of FET

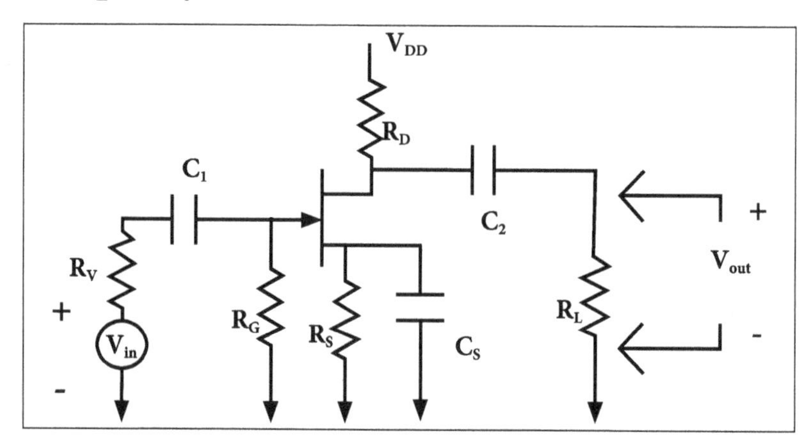

The Field Effect Transistor circuit low frequency response can be evaluated by analyzing the transfer functions of the elements affects the response at frequencies below mid-band. If there is more than one transfer function, the resultant overall response may be determined from the product of the individual response; generally done graphically with a Bode plot.

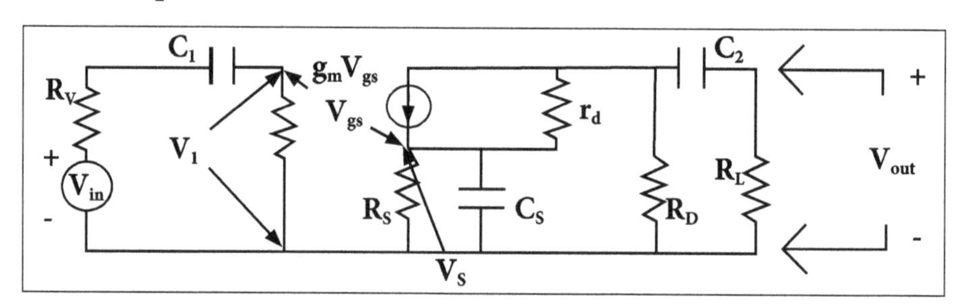

Figure shows the low frequency equivalent of the circuit of FET configuration in the previous figure. When examining the input circuit, the input transfer function is developed from C_1, R_G, R_V.

We know that,

$$V_1 = V_{in} R_G / (R_V + R_G) + (1 / j\omega C_1)$$

On simplifying,

$$V_1 / V_{in} = R_G / R_V + R_G \left(1 / \left(1 - j\left(1 / \omega (R_V + R_G) C\right)\right)\right)$$

$$V_1 / V_{in} = R_G / R_V + R_G \left(1 / \left(1 - j\left(\omega_1 (\text{input})\right) / \omega\right)\right)$$

$$(R_V + R_G) C = 1 / \omega_1 (\text{input}) = 1 / 2\, \pi\, f_1 (\text{input})$$

This expression reveals a pole in the low frequency response produced by the input elements C_1, R_G, and R_V at a frequency value of $f_1(\text{input}) = 1/2\pi(R_V + R_G)\,C_1$

The output elements of the circuit may also be expected to produce a low frequency pole. Evaluating the affect separately by assuming R_S in parallel with $C_S = Z_S = 0$. And r_d in parallel with $R_D = R$ the following equivalent circuit in the new figure can be produced.

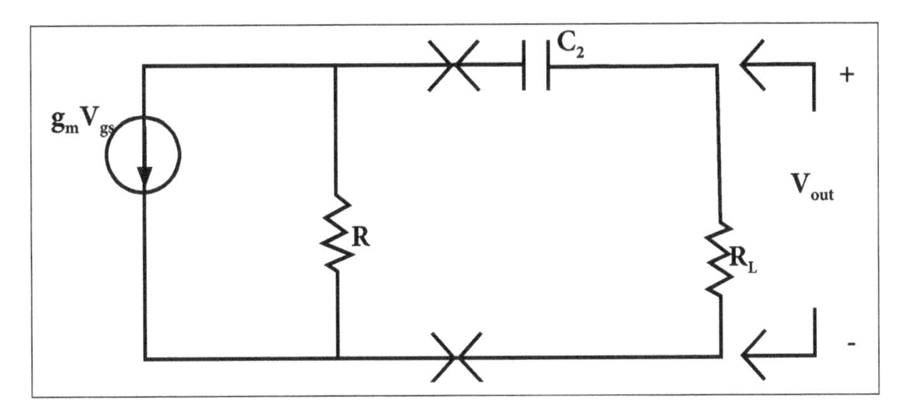

Applying a Thevenin's solution to the left of X - X we can redraw the equivalent circuit in the following figure:

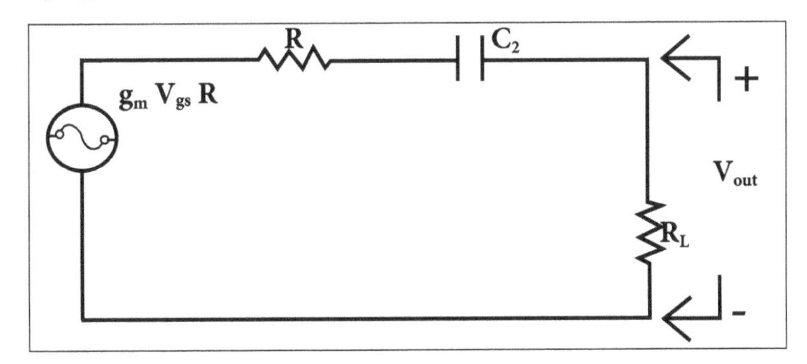

Therefore, the transfer function of Figure 4 becomes the following:

$$V_{üüüüüü}/V = V \;/g\; V\; R = R\; /R+R\left[1-(j/\omega(R+R\;)C\;\right]$$

$$üüüüüüü= \;_L \quad + \;_L\left[\quad -\left(\;\left(\omega_{1output} \quad \omega\right)\right)\right]$$

Now the transfer function becomes:

$$\omega_{1\ output} = 1/(R+R_L)C_2$$

Notice that this transfer function reveals another Pole in the low frequency response, at a frequency determined by the output coupling capacitor C_2 and the resistance the capacitor sees is $(R + R_L)$. Now, by disregarding the effects of R_S and C_S, the low frequency response looks like the following figure:

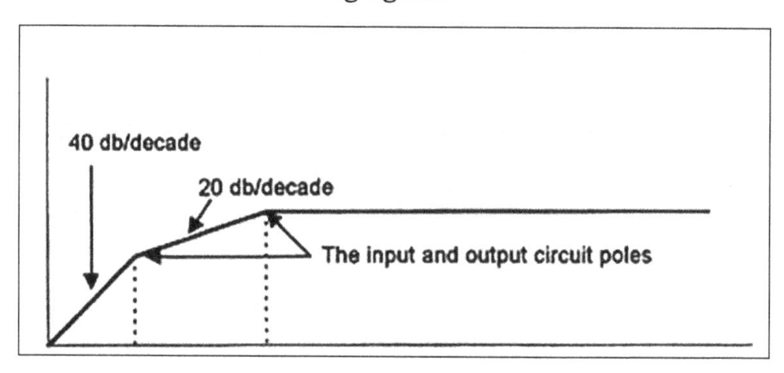

C_S in association with R_S must be assumed to also affect the low frequency response. It is possible to select values of R_S and C_S that remove their effects sufficiently to allow C_1 and C_2 to establish dominant poles. It is important to note that this case is not the same always. Recall the gain equation of the drain loaded amplifier with source impedance:

$$Av = -g_m Z_L / 1 + Z_S = -Z_L / (1/g_m) + Z_S$$

Effects of the source resistance R_S, and the source capacitance C_S are being considered and the load impedance Z_L is assumed to be resistive.

$$Av = -R / (1/g_m) + Z_S$$

And,

$$Z_S = R_S \text{ in Parallel with } C_S$$

$$Z_S = \left[R_S (1/j\omega C_S)\right] / \left[R_S + (1/j\omega C_S)\right]$$

$Z_S = R_S / (1 + j\omega R_S C_S)$, Substituting Z_S into the derived gain equation produces:

$$Av = -R / \left[1 / g_m + \left(R_S / \left(1 + (1 / j\omega R_S C_S)\right)\right)\right]$$

Rearranging the derivation with some general algebraic relations produces complete transfer function:

$$Av = -\left[R/(1/g_m + R_S)\right] \cdot \left[(1 + j\omega R_S C_S)/(1 + j\omega (R_S 1/g_m C_S)/(R_S + (1/g_m)))\right]$$

Examination of the completed transfer function result shows a zero at a frequency denoted with the expression of:

$$\omega_{So} = 1 / R_S C_S$$

And a pole at a frequency denoted as:

$$\omega_{Sp} = 1/(R_{S1}/G_m C_S)/(R_S + 1\, g_m)$$

The frequency response plot of the derived transfer function is represented graphically as shown in the below figure:

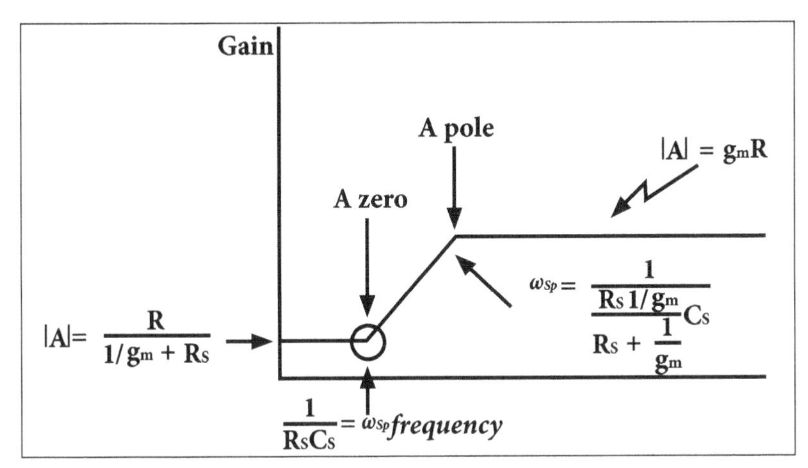

5.4 FET as an Amplifier

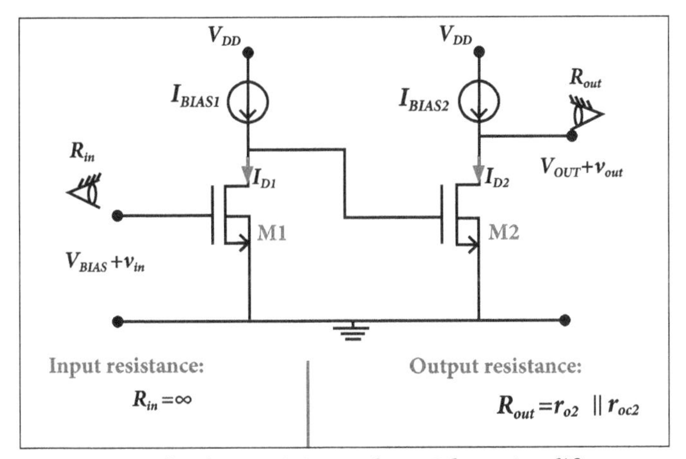

A Cascade of Two CS Stages for a Voltage Amplifier.

Input resistance:

$$R_{in} = \infty$$

Output resistance:

$$R_{out} = r_{o2} \,||\, r_{oc2}$$

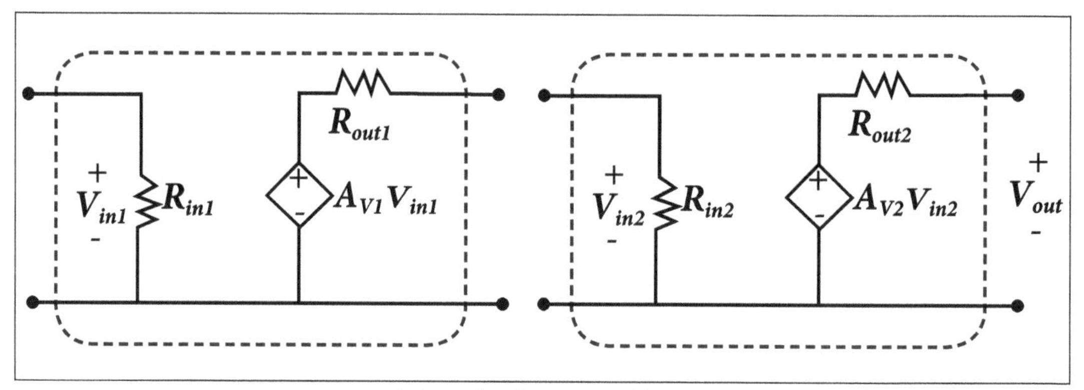

A Cascade of Two Amplifiers.

$$v_{out} = A_{v2} v_{in2} = A_{v2} A_{v1} v_{in1} \frac{R_{in2}}{R_{out1} + R_{in2}} = A_{v2} A_{v1} v_{in} \frac{R_{in2}}{R_{out1} + R_{in2}}$$

$$\Rightarrow A_v = \frac{v_{out}}{v_{in}} = A_{v2} A_{v1} \frac{R_{in2}}{R_{out1} + R_{in2}}$$

$$\begin{cases} \text{For the FET stages:} \\ R_{in1} = R_{in2} = \infty \end{cases}$$

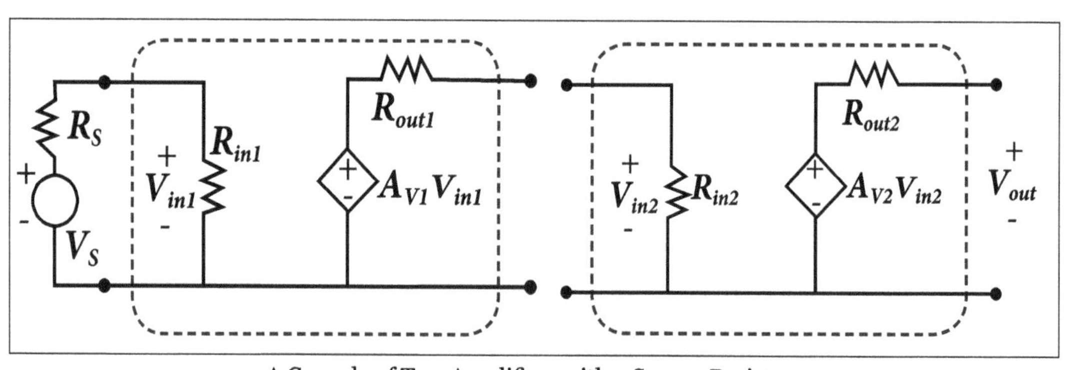

A Cascade of Two Amplifiers with a Source Resistor.

$$v_{out} = A_{v2} v_{in2} = A_{v2} A_{v1} v_{in1} \frac{R_{in2}}{R_{out1} + R_{in2}} = A_{v2} A_{v1} v_{in} \frac{R_{in2}}{R_{out1} + R_{in2}}$$

$$v_{out} = A_{v2} A_{v1} v_s \frac{R_{in}}{R_s + R_{in}} \frac{R_{in2}}{R_{out1} + R_{in2}}$$

$$\Rightarrow A_v = \frac{v_{out}}{v_s} = A_{v2} A_{v1} \underbrace{\frac{R_{in}}{R_s + R_{in}}}_{\text{Input voltage divider}} \underbrace{\frac{R_{in2}}{R_{out1} + R_{in2}}}_{\text{Inter-stage voltage divider}}$$

$$\begin{cases} \text{For the FET stages} \\ R_{in1} = R_{in2} = \infty \end{cases}$$

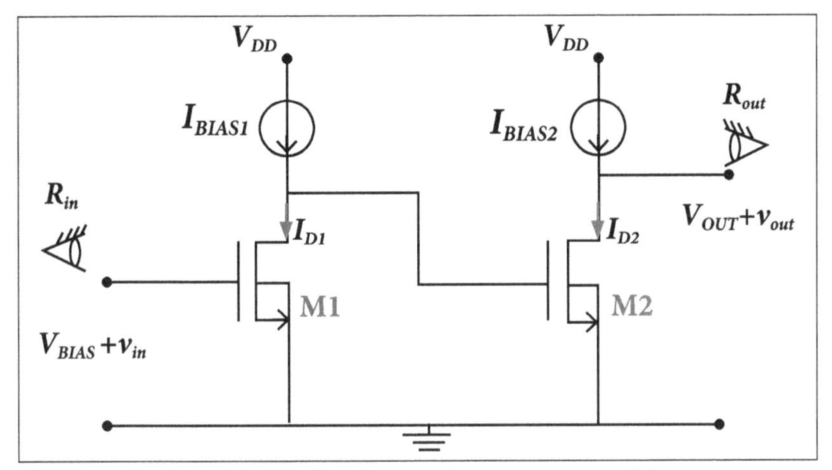

A Cascade of Two CS Stages for a Voltage Amplifier.

Input resistance:

$$R_{in} = \infty$$

Output resistance:

$$R_{out} = r_{o2} || r_{oc2}$$

Open circuit voltage gain:

$$A_v = \frac{v_{out}}{v_{in}} = A_{v1}A_{v2}$$

$$= \left[-g_{m1}\left(r_{o1} || r_{oc1}\right)\right]\left[-g_{m2}\left(r_{o2} || r_{oc2}\right)\right]$$

It is not a really a good voltage amplifier, if its output resistance is too large but it is a decent trans conductance amplifier.

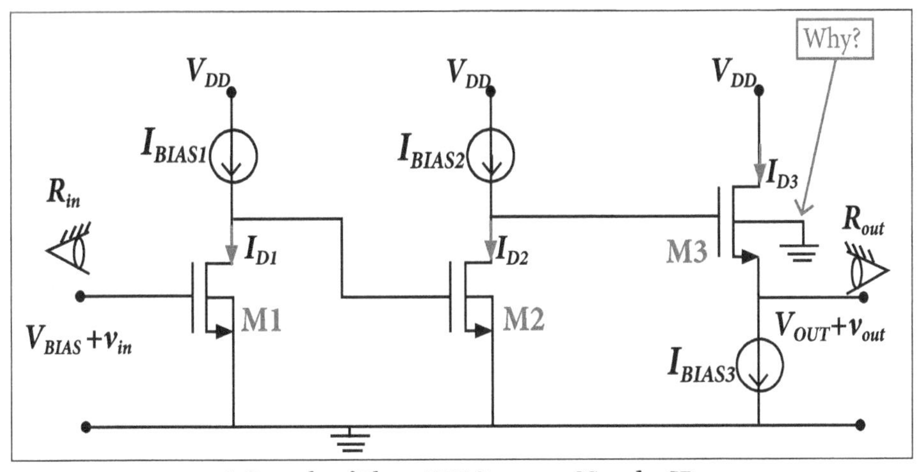

A Cascade of Three FET Stages: 2 CS and 1 CD.

Input resistance:

$$R_{in} = \infty$$

Output resistance:

$$\underbrace{\frac{1}{R_{out}}}_{\text{Small}} \approx \frac{1}{r_{oc3}} + \left(g_{m3} + g_{mb3}\right) \approx \left(g_{m3} + g_{mb3}\right)$$

Open circuit voltage gain:

$$\begin{aligned} A_v &= A_{v1} A_{v2} A_{v3} \\ &= \left[-g_{m1}\left(r_{o1} \,||\, r_{oc1}\right)\right]\left[-g_{m2}\left(r_{o2} \,||\, r_{oc2}\right)\right]\left[1\right] \end{aligned}$$

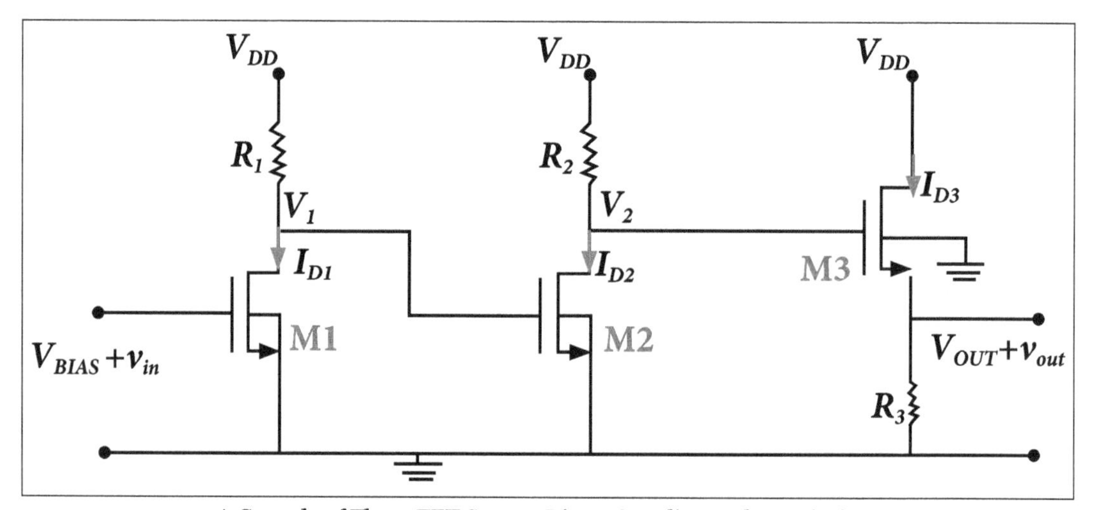

A Cascade of Three FET Stages: Direct Coupling and DC Biasing.

In a direct-coupled scheme, the DC bias of one stage affects, DC bias of other stages ,It is necessary to ensure appropriate DC bias of every stage such that:

- The desired voltage swing does not cause problems.
- The FETs are operating in saturation.

6

Amplifiers and Oscillators

6.1 Feedback Amplifiers

Feedback Systems process signals and same as signal processors. The processing part of a feedback system may be electrical or electronic, ranging from simple to complex circuits.

Simple analogue feedback control circuits are constructed using individual or the discrete components, like transistors, resistors and capacitors, etc, or even by using microprocessor-based and integrated circuits to form more complex digital feedback systems.

Open-loop systems are generally open ended and attempts are not made to compensate in order to changes circuit conditions or changes load conditions, due to variations in the circuit parameters, namely gain and stability, temperature, supply voltage variations and external disturbances. But the effects of the "open-loop" variations can be reduced when feedback is introduced.

A feedback system is one in which the output signal is sampled and then fed back to the input to form an error signal that drives the system. In general, Feedback is comprised of a sub-circuit that allows a part of the output signal from a system to modify effective input signal in a manner so as to produce a response that can differ substantially from that of the response produced in the absence of feedback.

Feedback Systems are very essential and widely used in amplifier circuits, process control systems, oscillators as well as various other types of electronic systems. The basic model of a feedback system is shown:

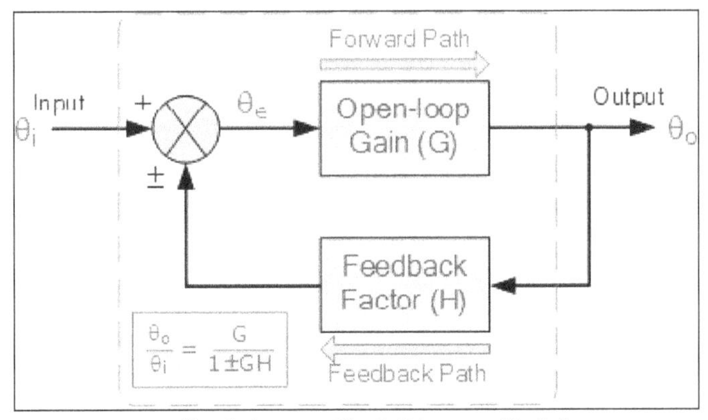

Feedback System Block Diagram Model.

This basic feedback loop of sensing, controlling and actuation is the main concept behind a feedback control system and there are various good reasons why feedback is applied and used in electronic circuits.

Circuit characteristics like system gain and response may be controlled. Circuit characteristics can be made independent of operating conditions such as supply voltages or variation in temperature.

The occurrence of signal distortion due to the non-linear nature of the components used can be reduced greatly. The Frequency Response, Gain and Bandwidth of a circuit can be controlled easily.

Mixing	sampling
Series → FB is v_f → I/P is v_s	Shunt → o/p is v_{out}
$V_f = \beta_{FB}\, v_{out}$ & This amplifier is Voltage Amplifier	
Series → FB is v_f → I/P is v_s	Series → o/p is i_{out}
$V_f = \beta_{FB}\, i_{out}$ & This amplifier is Transconductance Amplifier	
Shunt → FB is i_f → I/P is i_s	Series → o/p is i_{out}
$i_f = \beta_{FB}\, i_{out}$ & This amplifier is Current Amplifier	
Shunt → FB is i_f → I/P is i_s	Shunt → o/p is v_{out}
$i_f = \beta_{FB}\, v_{out}$ & This amplifier is Transresistance Amplifier	

Classification of Feedback Amplifiers

Types of feedback systems:

- Series-Shunt feedback.
- Series-Series feedback.
- Shunt-series feedback.
- Shunt-shunt feedback.

Feedback Concept, Transfer Gain

There are two main types of feedback control system which are:

- Negative Feedback.
- Positive Feedback.

Positive Feedback Systems

In positive feedback control system, the set point and output values are added together by means of controller as feedback is in-phase with the input. The effect of the positive feedback results in the improvement of the systems gain. System gain is the overall gain with positive feedback applied will be greater than the gain without feedback.

However, in electronic and control systems too much praise and the positive feedback will increase the systems gain too much, that would give rise to the oscillatory circuit responses as this increases the magnitude of the effective input signal.

An example of positive feedback systems can be an electronic amplifier based on an operational amplifier (op-amp) as shown in the figure:

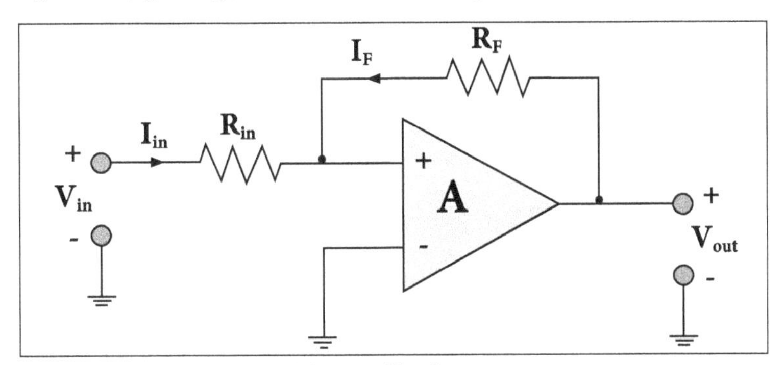

Positive Feedback System.

The Positive feedback control of the op-amp is achieved by applying a small part of the output voltage signal at V_{out} back to non-inverting (+) input terminal via the feedback resistor, R_F.

If the input voltage V_{in} is positive, the op-amp amplifies this positive signal and the output becomes positive. Some of this output voltage is returned back to the input by feedback network.

Thus the input voltage becomes more positive, causing an even larger output voltage and likewise, if the input voltage V_{in} is negative, the reverse happens and op-amp saturates at its negative supply rail. The positive feedback does not allow the circuit to function as an amplifier as the output voltage quickly saturates to one supply rail or the other, because with the positive feedback loops "more leads to more" and "less leads to less".

When the loop gain is positive, for any system then the transfer function is given as:

$$Av = G/(1-GH).$$

If GH = 1 the system gain, Av = infinity and the circuit will start to self-oscillate, after which no input signal is needed to maintain the oscillations, which is useful if an oscillator has to be made.

Although often considered as undesirable, this behavior is used in the electronics to determine a very fast switching response to a condition or signal. One example of the use of positive feedback is the hysteresis in which a logic device or system maintains a given state until a preset threshold is crossed by input. This type of behavior is termed as "bi-stability" and is often associated with the logic gates and digital switching devices like multi vibrators.

Positive or regenerative feedback increases gain and also the possibility of instability in a system that may lead to the self-oscillation. The positive feedback is widely used in oscillatory circuits such as Oscillators and Timing circuits.

6.2 General Characteristics of Negative Feedback Amplifiers

Negative Feedback Systems

In negative feedback control system output values are subtracted from each other since the feedback is out of phase with the original input. The effect of negative feedback is to decrease the gain. This is because the negative feedback produces stable circuit responses, improves stability and increases the operating bandwidth of a given system, the majority of all control causing feedback systems to be degenerative and reducing the effects of the gain.

An example of a negative feedback system is an electronic amplifier based on the functionality of an operational amplifier as shown in the figure:

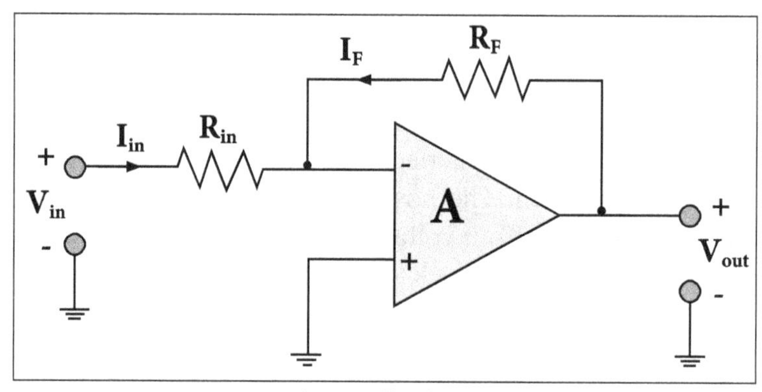

Negative Feedback System.

Negative feedback control of the amplifier is achieved by applying a small part of the output voltage signal at Vout back to inverting (–) input terminal via the feedback resistor, R_f.

If the input voltage V_{in} is positive, the op-amp amplifies this positive signal, but because it's connected to the inverting input of the amplifier and the output becomes more negative. Some of this output voltage is made to return back to the input by the feedback network of Rf.

Thus the input voltage is reduced by the negative feedback signal, that causes an even smaller output voltage and so on. Ultimately, the output will settle down and become stabilized at a value obtained by the gain ratio of R_f / R_{in}.

Same way, if the input voltage V_{in} is negative, then reverse may happen and the op-amps output becomes positive that adds to the negative input signal. Then we can see that negative feedback allows the circuit to function as an amplifier, so long as the output is within the limits of saturation. So we can see that the output voltage is stabilized and it is controlled by the feedback, because with negative feedback loops "more leads to less" and "less leads to more".

If the loop gain is positive for any system then the transfer function is given as,

$$Av = G / (1 + GH).$$

The use of negative feedback in amplifier and process control systems is widespread because ,as a rule negative feedback systems are more stable compared to the positive feedback systems and a negative feedback system is considered to be stable if it does not oscillate by itself at any frequency except for the given circuit conditions.

Another added advantage is that the negative feedback makes control systems more immune to the random variations in component values and the inputs. As the negative feedback significantly modifies the operating characteristics of a system it must be used with care.

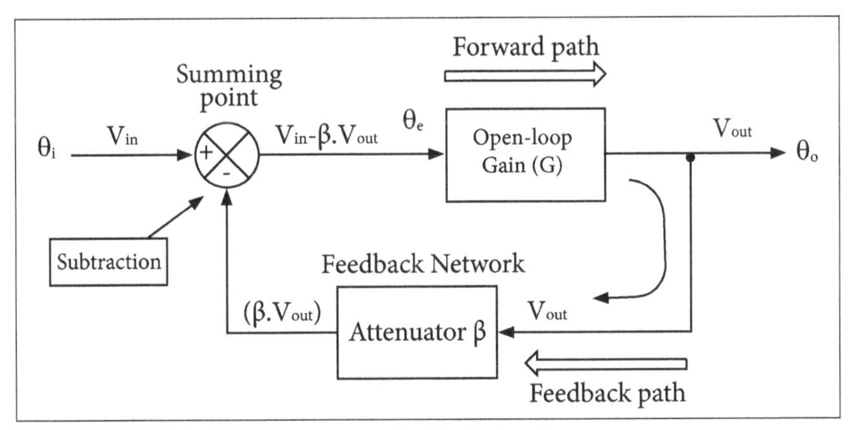

Negative feedback circuit.

The circuit represents a system with positive gain, G and feedback β. The summing junction at its input subtracts the feedback signal from the input signal to form the error signal V_{in} - βG, which drives the system.

Then using the basic closed-loop circuit above, the general feedback equation can be derived as,

System Gain,

$$G = \frac{V_{out}}{V_{in}}$$

$$G \times V_{in} = V_{out}$$

$$G(V_{in} - \beta V_{out}) = V_{out}$$

$$G.V_{in} - G.\beta.V_{out} = V_{out}$$

$$G.V_{in} = V_{out}(1 + G\beta)$$

$$\therefore \quad \frac{V_{out}}{V_{in}} = Gv = \frac{G}{1 + G\beta}$$

This is the negative feedback equation.

Where,

G = Open loop voltage gain

β = Feedback fraction

Gβ = Loop gain

1+Gβ = Feedback factor

Gv = Closed loop voltage gain

6.2.1 Effect of Feedback on Input and Output Resistances

The way that negative feedback is derived from the output of the amplifier and applied to the input will be used to modify the amplifier's input and output impedances so that impedance matching is maximized. For example an ideal voltage amplifier will have very high input impedance and very low output impedance; this will ensure that the maximum voltage waveform is passed from the previous circuit and transferred to the next circuit. Whereas, a current amplifier would need a very low output impedance to ensure that the maximum current is passed to the following circuit or output device.

6.3 Methods of Analysis of Feedback Amplifiers

Steps in Analyzing the Transistor Feedback Amplifiers:

- Identify the topology.
- Determine whether the feedback is positive or negative.
- Open the loop and calculate A, ß, R_i, and R_o.
- Use to find A_f, R_{if} and R_{of} or A_F, R_{iF}, and R_{oF}.
- Use the information in 4 to find whatever is required (v_{out}/v_{in}, R_{in}, R_{out}, etc.)

Generic Transistor Concept

Properties of a Generic Transistor

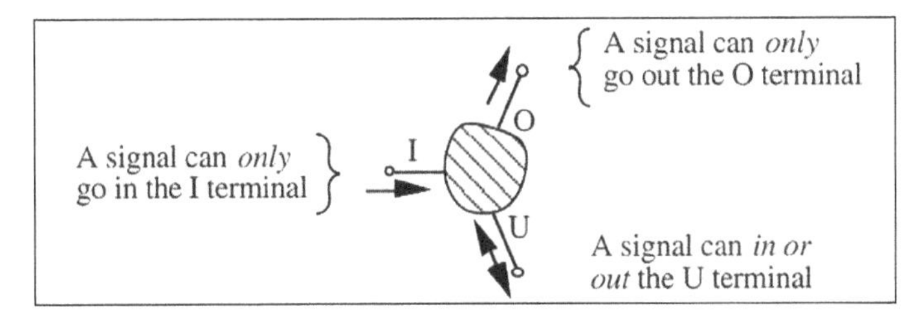

Identification of the Feedback Topology

Isolate the input and output transistors and apply the following identification:

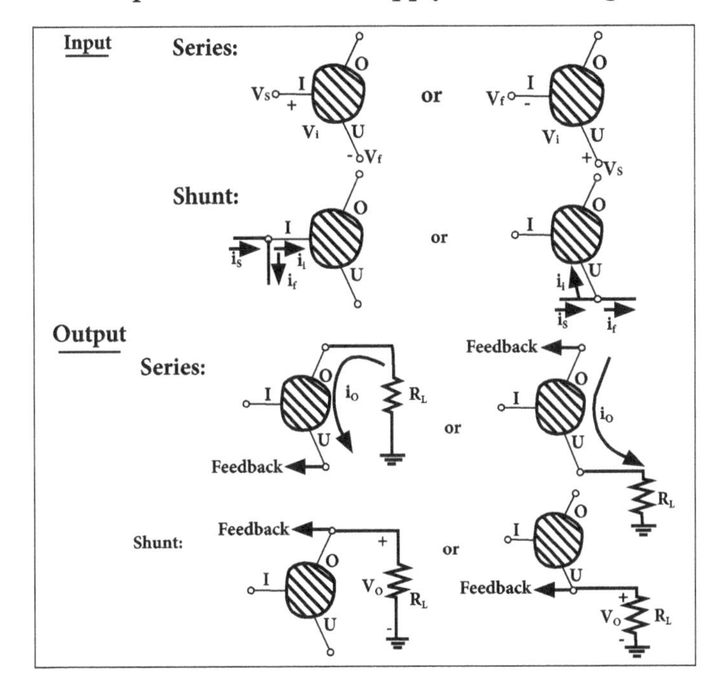

6.4 Power Amplifiers

Power amplifier refers to the amount of power provided by the power supply circuit or the amount of power delivered to the load. It is usually used in the last output stages of a circuit. The various examples of the power amplifiers include audio power amplifiers, servo motor controllers, push-pull amplifiers and RF power amplifiers.

Classification

Types of Power Amplifiers

The various types of power amplifiers are:

- Class A
- Class B
- Class AB
- Class C

Push-Pull Amplifiers

Class A power amplifier is a type of power amplifier where the output transistor is ON full time and the output current flows for the entire cycle of the input wave form.

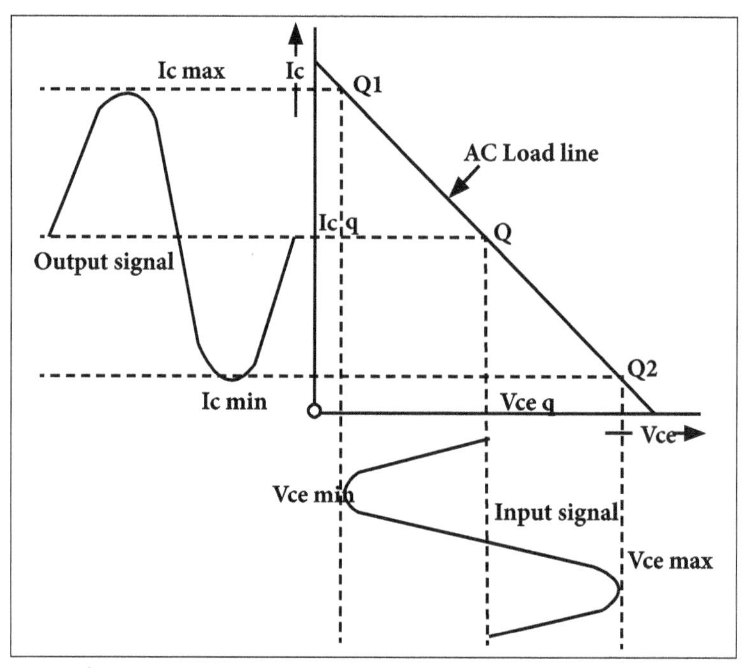

Class A power amplifier output characteristics-Ac load line.

Class A power amplifier is the simplest of all power amplifier configurations. They possess high fidelity and are totally immune to the crossover distortion but they have poor efficiency. Since the active elements are forward biased full time, some current will flow through them even when there is no input signal and this is the main reason for the inefficiency. Output characteristics of a Class A power amplifier is shown in the below figure.

From the below figure its is clear that the Q-point is placed exactly at the center of the DC load line and the transistor conducts for every point in the input waveform. The theoretical maximum efficiency of a Class A power amplifier is 50%.

In practical, with capacitive coupling and the inductive loads, the efficiency will be 25%. This means that 75% of the power used by the amplifier from supply line is being wasted. Majority of the power wasted is lost as heat on active elements. As a result, even a moderately powered Class A power amplifier requires a large power supply and a large heat sink.

Class A power Amplifier Circuit

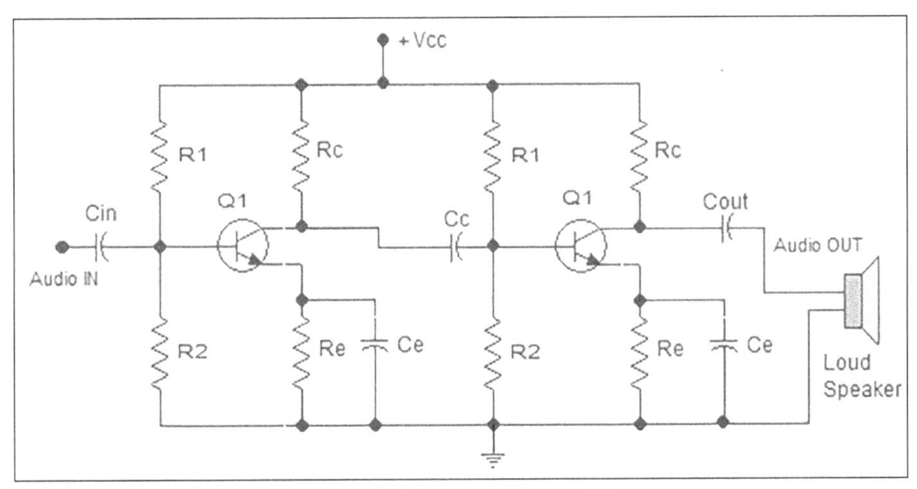

Two stage class A power Amplifier.

The circuit diagram of two stage single ended Class A power amplifier is shown. R1 and R2 are biasing resistors, which forms the voltage divider network that supplies the base of the transistor having a voltage 0.7V higher than the "negative maximum amplitude swing" of the input signal. This is the main reason behind the transistor being ON irrespective of the input signal amplitude. Capacitor C_{in} is the input decoupling capacitor that eliminates the DC components present in the input signal. If C_{in} is not present and there are DC components in the input signal, then the DC components will be directly coupled to the base of transistor and will surely alter the biasing conditions.

R_c is collector resistor and Re is emitter resistance. Their value is selected so that collector current is in desired level and the operating point is placed at the center of the load line under the zero signal condition. Placing the operating point as close as possible to the center of load line is very important for the distortion free operation of the amplifier. C_c is the coupling capacitor that connects the two stages. Its function is to block passage of DC components from 1st stage to the 2nd stage.

C_e is the emitter by-pass capacitor and its function is to by-pass the AC components in the emitter current while amplifier is operating. If C_e is not present, then the AC components will drop across the emitter resistor resulting in reduced gain. The most simple explanation is that, the additional voltage drop across R_e will get added to the base-emitter voltage and this means additional forward voltage is required to forward bias the transistor.

The output coupling capacitor C_{out} which couples output to the load. C_{out} blocks the DC components of the second stage from entering the load. The Coupling capacitor C_{out}, C_{in} and C_c all degrades the low frequency response of the amplifier. This is because these capacitors form high pass filters in conjunction with the input impedance of succeeding stages resulting in the attenuation of low frequency components. Input and output waveforms of a two stage RC couple amplifier is shown in the figure:

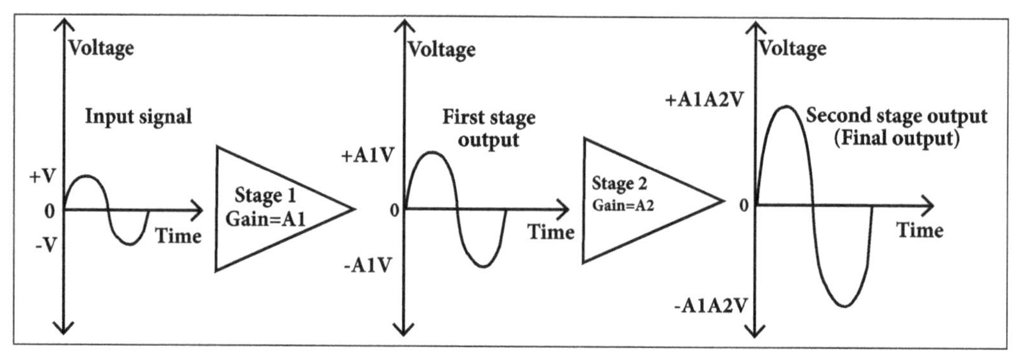

Two stages class A power amplifier input and output waveform.

Advantages

- High fidelity.
- Since the active device is on full time, no time is required for the turn on and this improves high frequency response.
- Single ended configuration can be practically realized in Class A amplifier. Single ended means only one active device in the output stage.
- Design is simple.
- Since the active device conducts for the entire cycle of the input signal, there will be no cross over distortion.

Disadvantages

- Poor efficiency.

- Steps for improving efficiency like the transformer coupling etc affects the frequency response.
- Powerful Class A power amplifiers are bulky and costly due to the large power supply and heat sink.

Transformer Coupled Class A Power Amplifier

An amplifier where the load is coupled with the output using a transformer is called a transformer coupled amplifier.

Using transformer coupling efficiency of the amplifier is improved.

The coupling transformer provides good impedance matching between the output and load .It is the main reason behind the improved efficiency. Impedance matching is a process of making the output impedance of the amplifier equal to the input impedance of the load which is an important criteria for the transfer of maximum power.

Circuit diagram of typical single stage Class A amplifier is shown in the circuit diagram:

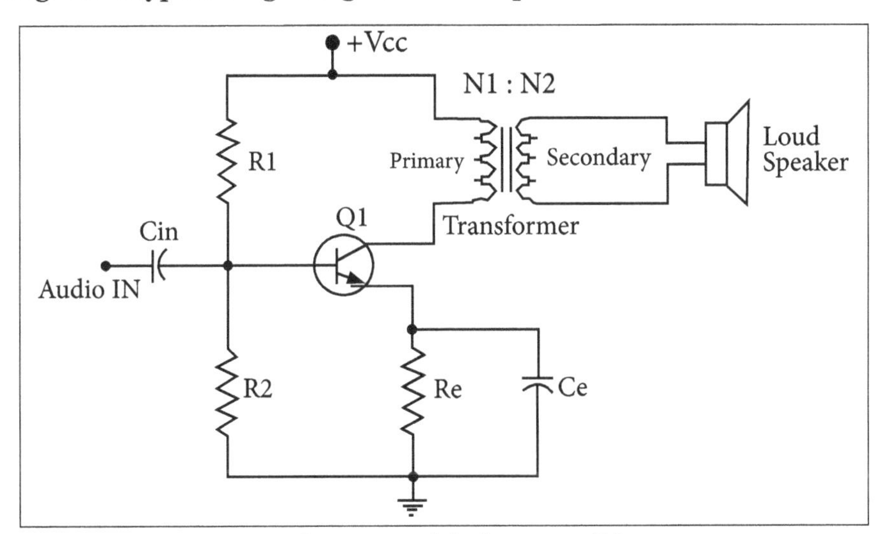

Transformer Coupled Class A Amplifier.

Impedance matching is achieved by means of selecting the number of turns of the primary so that the net impedance is equal to the output impedance of the transistors and by selecting the number of turns of the secondary so that its net impedance is equal to the input impedance loudspeakers.

Advantages

- High efficiency.
- Provides good DC isolation as there is no physical connection between amplifier output and load.

Disadvantages

- It is complex to find exactly matching transformer.
- Transformers are bulky and so it increases the cost and size of the amplifier.
- Transformer winding will not provide any resistance to DC current. If any DC components if present in the amplifier output, it will flow through primary winding and saturate the core. This will result in reduced transformer action.
- It can be used only for small loads.
- It reduces the low frequency response of the amplifier

6.5 Harmonics Distortion Factor

Harmonic Distortion Calculations

Harmonic distortion (D) can be calculated:

$$\%\text{nth harmonic distortion} = \%D_n = \left|\frac{A_n}{A_1}\right| \times 100$$

Where A_1 is the amplitude of the fundamental frequency A_n is the amplitude of the highest harmonic.

The total harmonic distortion (THD) is determined by:

$$\%THD = \sqrt{D_2^2 + D_3^2 + D_4^2 + \ldots} \times 100$$

6.5.1 Oscillators

Oscillators are classified depending upon the following parameters as given below:

- The wave shapes that is generated by the oscillators.
- The fundamental mechanisms used in the oscillators.
- Range of frequency of output signal.
- Type of circuit.

Based on the Wave Shapes Generated by the Oscillators:

- Sinusoidal Oscillators: Sinusoidal oscillator generates a sine-wave output signal.
- Non-sinusoidal Oscillators: Non-sinusoidal oscillator generates non-sinusoidal or complex waveforms such as square, rectangular, saw tooth and trapezoidal.

Based on the Operating Frequency

Types of Oscillators	Frequency Range	Example
Audio frequency oscillators (AFO)	400 Hz to 20 kHz	RC oscillators such as Phase-shift and Wien-bridge oscillators.
Radio frequency oscillators (RFO)	20 kHz to 30MHz	LC feedback oscillators such as tuned collector, Hartley and Colpitt oscillators.
Very high frequency (VHF) oscillators	30 MHz to 300 MHz	Crystal oscillators used in microprocessors, microcontrollers, ASICs, DSP processors and computer mother board.
Ultra high frequency (UHF) oscillators	300 MHz to 3 GHz	Bulk Acoustic Wave (BAW) AT cut quartz crystal oscillators.
Microwave frequency oscillators	Above 3 GHz	YIG tuned oscillators, oven controlled crystal oscillators.

Based on the Circuit Components

Oscillators are also be classified according to type of circuit as given below:

- LC tuned oscillators.
- RC phase-shift oscillators.

6.5.2 Condition for Oscillation

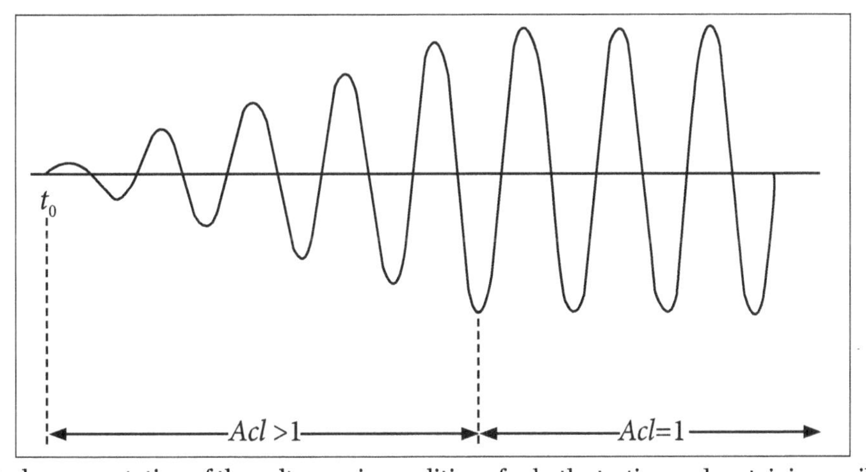

Graphical representation of the voltage gain conditions for both starting and sustaining oscillation.

The above graph shows that, when oscillation starts at to, the condition $A_{cl} > 1$ is the one that causes the sinusoidal output voltage amplitude to build up to a desired level. Then A_{cl} decreases to 1 and maintains desired amplitude.

The starting voltage is provided by noise, which is produced by means of random motion of electrons in resistors used in the circuit. The noise voltage contains almost all the sinusoidal frequencies. This low amplitude noise voltage gets amplified and appears at the output terminals.

The amplified noise drives the feedback network which is a phase shift network. As a result of this feedback voltage is maximum at a particular frequency, which in turn represents the frequency of oscillation.

6.5.3 RC-Phase Shift Oscillator and Wein Bridge Oscillator

RC phase shift oscillators is an audio frequency oscillator or low frequency oscillator. It uses a common emitter (CE) amplifier whose output is given to the three RC networks. The phase shift produced by the common emitter amplifier is 180°.

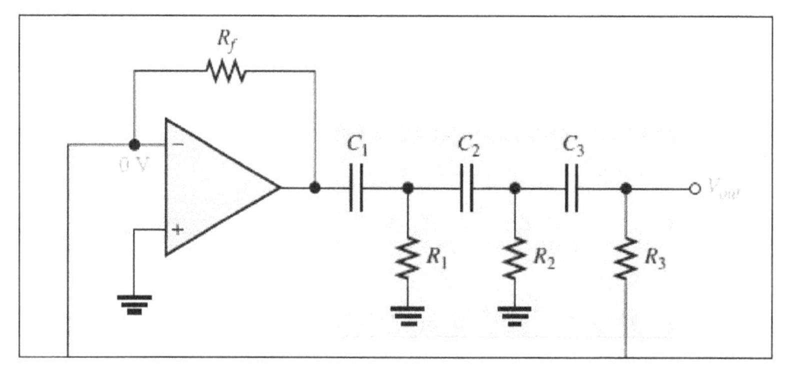

Diagrammatic representation of RC phase shift oscillators.

As an oscillator requires a phase shift of 0° or 360° and the additional 180° phase shift is obtained using three RC networks with an individual phase shift of 60° each.

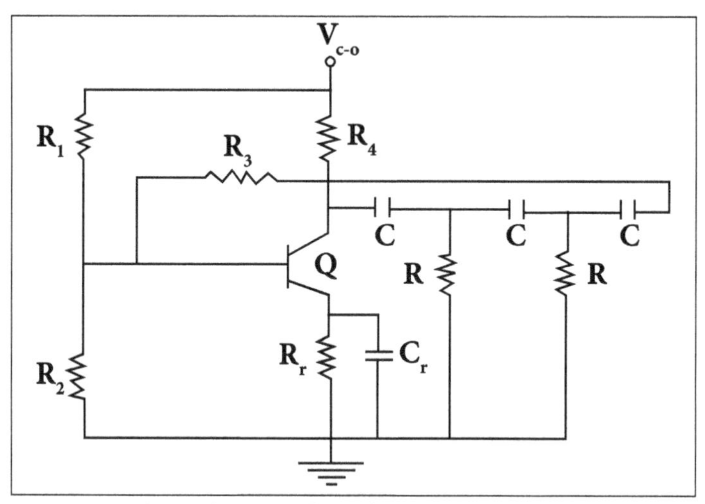

Circuit diagram of RC phase shift oscillator.

Circuit Operation

The circuit starts operating if there exists any inherent noise signal in the transistor or any variations in the power supply. With this input at the base, the amplifier produces a collector current.

The voltage at the collector is amplified and shifted by 180° in phase. This is fed to the RC network which shifts the phase by 180° and feeds the signal in phase with the base current. This increases the base bias.

Wien Bridge Oscillators with BJT and FET and their Analysis

The Wien-Bridge oscillator is also an audio frequency oscillator. It is the only one oscillator which involves both positive and negative feedback. Negative feedback provides stability and positive feedback provides oscillation.

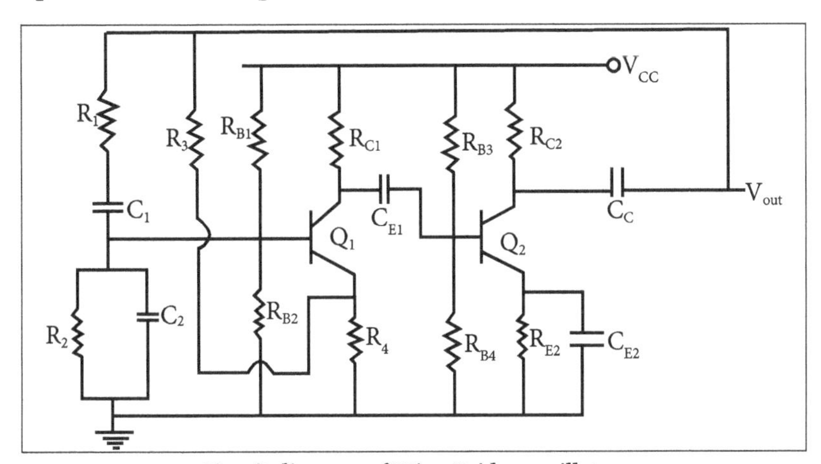

Circuit diagram of Wien Bridge oscillator.

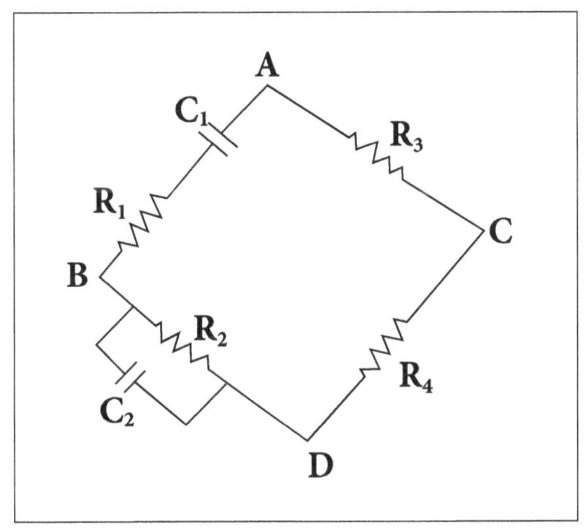

Circuit representation of Bridge circuit.

In the bridge circuit, a two stage CE transistor amplifier is used. Each stage contributes a 180° phase shift thus the total phase shift due to an amplifier stage becomes 360°. Hence feedback network produces no phase shift.

Though the signal at the output of Q_2 is in phase with the input at the base of it cannot be directly fed as a feedback signal, as it would affect the frequency stability.

Hence a Wien-Bridge circuit which is sensitive to only one voltage of a particular frequency is added, through which the output voltage of Q_2 is coupled to the input of Q_1. This increases the frequency stability since Q_1 is made to respond to only one frequency.

The bridge circuit consists of two arms.

6.5.4 Crystal Oscillator

Miller Oscillator

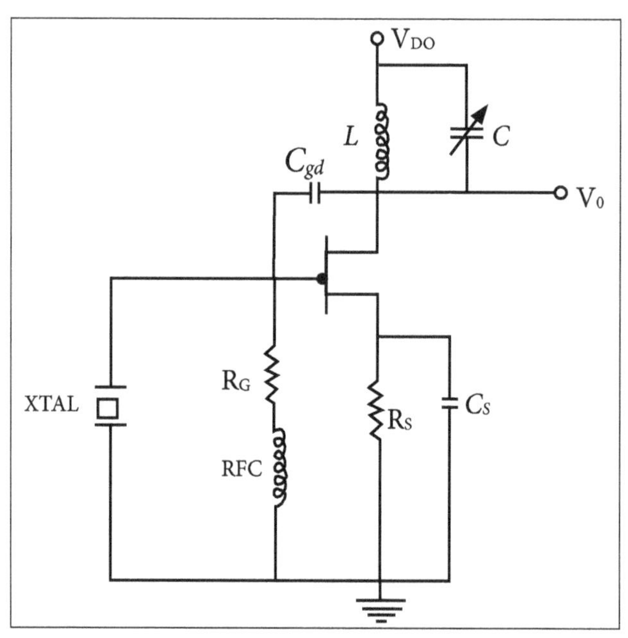

Circuit diagram of Miller Crystal Oscillator.

Miller Crystal Oscillator is one of the crystal controlled oscillator circuit. The crystal is operated in its parallel resonant mode where it offers maximum impedance. The crystal is connected across gate to source of the FET amplifier. The tank circuit is connected at the drain terminal.

At the resonant frequency of crystal, the gate to source bias is minimum so the drain current is maximum. Thus, the LC tank circuit is set into sustained oscillations. At frequencies other than resonant frequencies, the gate to source voltage is maximum and therefore the drain current is low which results in the damped oscillations. Hence, the circuit results in undammed or sustained oscillations only at the crystal resonant frequency which results in a better frequency stability. The frequency of oscillation is determined by the LC value of the tank circuit.

In the above circuit, the crystal forms one arm and the tank circuit forms the other arm.

The inter junction capacitance C_{gd} acts as capacitor between D and G. Since the circuit involves the effect of C_{gd}, it is termed as the Miller Oscillator.

Pierce Crystal Oscillator

Oscillators are electronic circuits that generate an output signal with the necessity of an input signal. In order to excite a crystal for operation in the series resonant mode it can be connected as a series element in the feedback path. At the series resonant frequency of the crystal, its impedance is the smallest and amount of feedback is the largest. Resistors R_1, R_2, R_E provide the voltage divider bias. C_E is the bypass capacitor and RFC coil provides for the dc bias while decoupling any AC signal on the power lines from affecting the output signal.

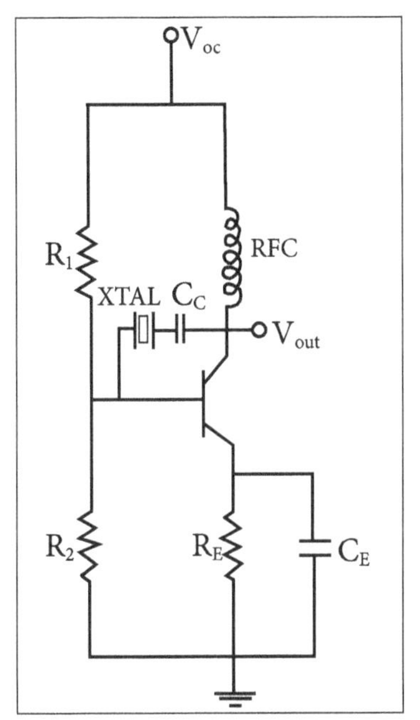

Circuit representation of Pierce crystal Oscillator.

The voltage feedback from the collector to base is a maximum when the crystal impedance is minimum. The coupling capacitance C_C has negligible impedance at the circuit operating frequency but blocks any dc between collector and base.

The resulting circuit frequency of oscillation is set, then by the series resonant frequency of crystal. Changes in the supply voltage, transistor device parameters and so on these have no effect on the circuit operating frequency which is held stabilized by the crystal. The circuit frequency stability is set by the crystal frequency which is good.

6.5.5 Frequency and Amplitude Stability of Oscillators

Even if an oscillator is set at an initial frequency, it is not maintained throughout. They keep on changing. The term "frequency stability" is used to define the ability of the oscillator to maintain a single fixed frequency as long as possible over a time interval. The deviations in frequency are caused due to variations in the values of circuit features that determine the oscillator frequency.

Other factors responsible for drift in oscillator frequency are given below:

Operating Point of Active Device

The effects of variations in inter element capacitances can be neutralized by introducing a swamping capacitor across the offending elements.

Inter-element Capacitances

If the operating point of the active device in the circuit is in the non-linear portion of its characteristics, there may be variations in the transistor parameters which in turn affect the oscillator frequency stability. So that the operating point, Q is carefully selected to work in the linear portion of characteristics of the active device.

Mechanical Vibrations

Although the mechanical vibration is not such a high frequency stability changing factor, they can be easily avoided by isolating the oscillator circuit from source of the mechanical vibrations.

When the circuit operates for a long time, the heat starts to build up. As a result, the values of the frequency determining components like resistors, inductors and capacitors change with temperature. Thus transistor parameter values also tend to change. But, the change in the values of R, L and C will be slow and thus the change in oscillator frequency will also be slow.

The other major factor responsible for deviation in frequency is variations in power supply. However, this problem can be overcome by using regulated power supply. Any variation in load coupled to the tank circuit may cause change in effective resistance of the circuit by transformer action which in turn causes the drift in frequency.

Permissions

All chapters in this book are published with permission under the Creative Commons Attribution Share Alike License or equivalent. Every chapter published in this book has been scrutinized by our experts. Their significance has been extensively debated. The topics covered herein carry significant information for a comprehensive understanding. They may even be implemented as practical applications or may be referred to as a beginning point for further studies.

We would like to thank the editorial team for lending their expertise to make the book truly unique. They have played a crucial role in the development of this book. Without their invaluable contributions this book wouldn't have been possible. They have made vital efforts to compile up to date information on the varied aspects of this subject to make this book a valuable addition to the collection of many professionals and students.

This book was conceptualized with the vision of imparting up-to-date and integrated information in this field. To ensure the same, a matchless editorial board was set up. Every individual on the board went through rigorous rounds of assessment to prove their worth. After which they invested a large part of their time researching and compiling the most relevant data for our readers.

The editorial board has been involved in producing this book since its inception. They have spent rigorous hours researching and exploring the diverse topics which have resulted in the successful publishing of this book. They have passed on their knowledge of decades through this book. To expedite this challenging task, the publisher supported the team at every step. A small team of assistant editors was also appointed to further simplify the editing procedure and attain best results for the readers.

Apart from the editorial board, the designing team has also invested a significant amount of their time in understanding the subject and creating the most relevant covers. They scrutinized every image to scout for the most suitable representation of the subject and create an appropriate cover for the book.

The publishing team has been an ardent support to the editorial, designing and production team. Their endless efforts to recruit the best for this project, has resulted in the accomplishment of this book. They are a veteran in the field of academics and their pool of knowledge is as vast as their experience in printing. Their expertise and guidance has proved useful at every step. Their uncompromising quality standards have made this book an exceptional effort. Their encouragement from time to time has been an inspiration for everyone.

The publisher and the editorial board hope that this book will prove to be a valuable piece of knowledge for students, practitioners and scholars across the globe.

Index

Printed in the USA
CPSIA information can be obtained
at www.ICGtesting.com
JSHW060948081123
51533JS00033B/50

9 781639 897179